I0820714

GREAT DISCOVERIES IN SCIENCE

Semiconductors

Grace Murphy

New York

Published in 2018 by Cavendish Square Publishing, LLC
243 5th Avenue, Suite 136, New York, NY 10016

Copyright © 2018 by Cavendish Square Publishing, LLC

First Edition

No part of this publication may be reproduced, stored in a retrieval system, or transmitted in any form or by any means—electronic, mechanical, photocopying, recording, or otherwise—without the prior permission of the copyright owner. Request for permission should be addressed to Permissions, Cavendish Square Publishing, 243 5th Avenue, Suite 136, New York, NY 10016.
Tel (877) 980-4450; fax (877) 980-4454.

Website: cavendishsq.com

This publication represents the opinions and views of the author based on his or her personal experience, knowledge, and research. The information in this book serves as a general guide only. The author and publisher have used their best efforts in preparing this book and disclaim liability rising directly or indirectly from the use and application of this book.

CPSIA Compliance Information: Batch #CS17CSQ

All websites were available and accurate when this book was sent to press.

Library of Congress Cataloging-in-Publication Data

Names: Murphy, Grace (Nonfiction author)
Title: Semiconductors / Grace Murphy.
Description: New York : Cavendish Square Publishing, [2018] | Series: Great discoveries in science | Includes bibliographical references and index.
Identifiers: LCCN 2016058607 (print) | LCCN 2017003784 (ebook) | ISBN 9781502628718 (library bound) | ISBN 9781502628725 (E-book)
Subjects: LCSH: Semiconductors.
Classification: LCC TK3301 .M87 2018 (print) | LCC TK3301 (ebook) | DDC 537.6/22--dc23
LC record available at https://lccn.loc.gov/2016058607

Editorial Director: David McNamara
Editor: Caitlyn Miller
Copy Editor: Michele Suchomel-Casey
Associate Art Director: Amy Greenan
Designer: Lindsey Auten
Production Coordinator: Karol Szymczuk
Photo Research: J8 Media

The photographs in this book are used by permission and through the courtesy of: Photos credits: Cover Vinne/Shutterstock.com; p. 4 Victor R. Boswell Jr./National Geographic/Getty Images; p. 8 Hinterhaus Productions/Taxi/Getty Images; p. 11, 43 Bettmann/Getty Images; p. 13 Jan Schneckenhaus/Shutterstock.com; p. 18 Andrew Lambert Photography/Science Source; p. 21 Stapleton Historical Collection/Heritage Image Partnership Ltd/Alamy Stock Photo; p. 24 PhotoQuest/Archive Photos/Getty Images; p. 27 Science & Society Picture Library/SSPL/Getty Images; p. 29 Designua/Shutterstock.com; p. 30 Leemage/Universal Images Group/Getty Images; p. 34 Sergey Merkulov/Shutterstock.com; p. 35 José Antonio Peñas/Science Source; p. 38 Encyclopaedia Britannica/Universal Images Group/Getty Images; p. 45 Dorling Kindersley/Thinkstock; p. 50 Steve Larson/Denver Post/Getty Images; p. 52 Apic/Hulton Archive/Getty Images; p. 56 Robert W. Kelley/The LIFE Picture Collection/Getty Images; p. 62 Dan McCoy/The LIFE Images Collection/Getty Images; p. 68 Walter Zerla/Cultura/Getty Images; p. 71 Hulton Archive/Archive Photos/Getty Images; p. 79 generalfmv/iSTock/Thinkstock; p. 84 krystiannawrocki/E+/Getty Images; p. 88 Maxx-Studio/Shutterstock.com; p. 90 Everett Historical/Shutterstock.com; p. 92 Justin Sullivan/Getty Images; p. 102 Macrovector/Shutterstock.com.

Printed in the United States of America

Contents

An integrated circuit chip, made from a semiconductor, threads the eye of a needle.

Introduction: A Hidden Revolution

Semiconductors are everywhere. They are at the electronic heart of cell phones, laptops, airplanes, smartwatches, automobiles, and even many washing machines. We don't often think about them, and we may not even use the word "semiconductor." But without semiconductors, the world as we know it today would be a very different place.

The definition of "semiconductor" is very simple: "a material whose electrical conductivity is between that of a conductor and that of an insulator." That doesn't sound very dramatic. Yet the revolution that started when scientists began to understand the implications of that idea was far from simple, and so was harnessing the power of semiconductors. Even today, the massive power and endless uses of semiconductors are still being explored, understood, and applied.

The story of semiconductors begins with Michael Faraday, sitting in a laboratory in London, trying to understand how electricity works. Like Benjamin Franklin, in his famous experiment with the lightning rod, Faraday conducted investigations to help him understand the characteristics of different materials and how electric charges were transmitted. He discovered that one of the substances that he was working with, silver sulfide, had a peculiar property. Its ability to conduct electric charges increased when it was heated. Thus, the first semiconductor was identified.

Faraday's observation began an exciting journey, one that includes the discovery of the structure of the atom and moves into quantum physics and beyond. That journey, which is still unfolding, has resulted in an explosion of theoretical knowledge about the nature of the universe and practical applications of that knowledge in our everyday lives.

The journey takes us through the nineteenth century, when scientists observed more about the properties of semiconductors but still did not understand them. At the same time, other developments in the uses of **electrical currents** led to significant changes in society. The invention of the telegraph transformed communication. The **vacuum tube** became the basis of radioelectronics. Semiconductors, however, did not seem to have a place in these developments.

Yet some scientists continued to work with these mysterious materials. J. J. Thomson's identification of **electrons** as a component of atoms contributed to the understanding of semiconductors. In Russia, Abram F. Ioffe published a book on semiconductors. In the 1920s and 1930s, scientists continued to develop greater understanding of the fundamental physics of semiconductors. Vacuum tubes, until then at the center of radioelectronics, were showing their limitations—they were too slow, they were too large, and they burned out too quickly. Semiconductors finally showed their potential.

Semiconductors, and the information and communication revolution they created, first took concrete form at Bell Laboratories with the invention of the **transistor** in 1947. Then, in 1958 in Santa Clara Valley (soon to be called **Silicon** Valley), the **integrated circuit**, containing thousands of transistors in a single small wafer, was invented. Shortly thereafter, the **microprocessor**, with the ability to store information and process programmed instructions about how to handle that information, was created.

Tens and eventually hundreds of companies were founded that used and improved the new technologies. Computers became faster and smaller, and these machines, which originally occupied enormous rooms, fit first in a home and then finally in a pocket. Most devices we can think of use semiconductor technology to improve their capabilities and extend their usefulness. Cars, trains, and planes all use semiconductor chips. Microwaves, stoves, and refrigerators have semiconductor components. Digital cameras and GPS systems use semiconductor parts. And semiconductor-based cell phones have become the indispensable accessory for everything we do.

We may not talk much about semiconductors, but they have changed the way we can talk. Semiconductors have allowed us to communicate easily with people on the other side of Earth and even people in outer space. They have changed the form of messages we send from letters that took weeks or even months to transmit to words and images that can go around the globe in an instant.

The story of semiconductors is the story of our modern age, and it is a truly momentous tale.

Cell phones link us together today, but communication hasn't always been so simple.

CHAPTER 1

The Problem of Information

We live in the Information Age. We have access to endless amounts of information, and we have an enormous variety of ways to exchange that information. We can call a friend, watch television, download music, or read a book. But the ability to get new information and to share our knowledge wasn't always this easy. This chapter looks at how the transmission of information has changed over time and how the semiconductor came to play an integral role in that exchange.

EARLY INFORMATION EXCHANGE

From the beginning of human history, we have been trying to learn more about the world around us and communicate what we know to other human beings. Over time, we developed a variety of ways to accomplish this. Early peoples told stories about the world through the oral tradition—passing on information by speaking directly to one another. Some early cultures also used smoke signals to provide information at a distance.

People began to carve information on stone, using tools and developing alphabets to do this more efficiently. Letters were drawn on papyrus, an ancient form of paper, and on parchment or vellum, made from animal skins. But all these forms of communication had the same problem: they were limited in how much information they could provide and by how many people could get access to that information.

The PRINTING PRESS

In the fifteenth century, a German metalworker named Johannes Gutenberg introduced the first printing press to Europe. This was a momentous event, and it is considered to be the beginning of the modern age of history. With the ability to produce many copies of the same document or book (the first one was a Bible), the printing press made mass communication possible. The explosion of the exchange of information and ideas led to many societal changes.

The increase in literacy contributed to the rise of a middle class and an increase in education and learning. There were political revolutions against the ruling classes, reforms of the religious institutions, and the spread of cultural and artistic awareness. As information became more available to more people, scientific inquiry also became more frequent. Scientists began exploring the phenomenon that was to provide the basis for the next great information explosion: electricity.

OBSERVING ELECTRICITY: From STONES to LIGHTNING RODS

A Greek philosopher named Thales noted that amber was able to attract other light objects such as straw or dust when rubbed with wool. This early observation of static electricity didn't develop much further for centuries. Then, at the end of the sixteenth century, the information explosion that began with the printing press resulted in a renewed interest in electricity.

Johannes Gutenberg and the printing press

Johannes Gutenberg

Johannes Gutenberg, inventor of the printing press, was born in Mainz, Germany, at the end of the fourteenth century. His invention, which provided for the first mass communication method, ushered in the modern age.

Gutenberg was the son of a merchant family and was trained in metalwork. A harsh battle between the craftsman guilds in Mainz in 1428 led to the Gutenbergs' exile in Strasbourg, where Johannes began his printing press experiments. He had to borrow money to get the supplies for his secret invention, and much of the information about Gutenberg comes from local financial records.

When Gutenberg returned to Mainz in 1448 to run a printing business, he borrowed more money. This time it was from Johann Fust, a Mainz financier. Gutenberg was able to acquire the specific tools and supplies that he needed to continue the work on his invention, and Fust became a partner in the business. Gutenberg's invention was based on an Asian system of movable type but used his own metal casting system and oil-based ink.

Sometime around 1455, Gutenberg printed his first masterpiece, the Forty-Two Line Bible. At the same time, however, Fust became discouraged at the delay in any return on his investment and sued Gutenberg. Fust won his suit and gained control of the type for the Bible and for Gutenberg's second masterpiece, a Psalter. As a result, the products of Gutenberg's world-changing invention were printed under the name of another printer.

The first printing press

William Gilbert, a physician to Queen Elizabeth, began studying magnetic properties. His experiments demonstrated that many substances (including amber) attracted the needle of an instrument Gilbert had built. This device was a prototype of an **electroscope**, an instrument used to measure whether or not an object is electrically charged. Gilbert also proved that other substances did not have this attraction. He called this property of attraction "electrical," based on the Greek word for amber. Soon other scientists were performing electrical experiments, learning more about electrical conduction and identifying substances that carried electrical charges. A professor of physics and theology at Oxford University, J. T. Dedaguliers, named the substances that carried electricity conductors and those that did not insulators.

Electricity became a popular phenomenon. Electric demonstrations provided entertainment, and electrified water was thought to be good for your health. In America, Benjamin Franklin noted the similarity between the sparks produced by electrical machines and lightning. Franklin's experiment involved flying a kite when a storm was coming. He attached a metal key to the end of a rope close to the ground. Then, when lightning came and metallic objects were placed close to the key, electric sparks were emitted—just as they were in the electrical machines.

UNDERSTANDING ELECTRICITY: From BATTERIES to ELECTRONS

In Italy, Luigi Galvani conducted experiments on frogs. As his scalpel touched the clamp holding the frog in place, the amphibian's leg twitched. Galvani believed that this indicated the presence of what he thought of as animal electricity. His colleague, Alessandro Volta—who gave his name to the volt, a measure of electromotive force—realized that the response was not caused by the frog. Rather, it resulted when the two different metals in the scalpel and the clamp came in contact.

Volta also invented the voltaic pile. This early electric battery was composed of silver and zinc discs, separated by wet cardboard, and provided a reliable source of direct electric current. Volta's device stimulated considerable experimentation and discovery all over Europe. He was also the first scientist to use the term "semiconductor."

Hans Christian Øersted, a Danish physicist, conducted experiments demonstrating that electrical currents create magnetic fields. In France, Andre Marie Ampere provided a mathematical and physical theory to explain the relationship between electricity and magnetism. Although he is most famous for what came to be called **Ampere's law**, he also theorized that the **electromagnetic** connection was a result of the presence of an electromagnetic molecule. This idea foreshadowed the identification of the electron. In Germany, George Ohm mathematically described conduction in circuits. In 1821, another German physicist, Tomas Seebeck, created experiments in which he noted that heating the joint of two different metals created a current and that the strength of that current varied depending upon the materials used in the joints. These early investigations provided a basis for the use of semiconductors over a century later.

Then, in England, Michael Faraday, a self-educated physicist and chemist, made an unusual observation. Faraday knew that when the temperature increases, the conductivity of certain metals decreases. But while conducting an experiment with silver sulfide—considered to be a metal—he noted the conductivity of the substance *increased* as the temperature increased. He searched for other substances that possessed this peculiar characteristic, and over time, he found several other examples, including lead fluoride and mercuric sulfide. Faraday described these findings in his noted work, *Experimental Research in Electricity.* Although Faraday's work was well regarded and he was noted for the discovery of the laws of electromagnetic induction and **electrolysis**, his observations on the odd response of certain materials to heat did not inspire his colleagues to explore the phenomenon further. In fact, less than two decades after Faraday's experiments, Johann Hittorf, a German physicist, stated that there was nothing unusual in the heat-dependent conductivity of silver sulfide and that it did not indicate a different class of substance. Faraday, though, was right.

In 1839, Alexander Becquerel, a nineteen-year-old French physicist working in his father's laboratory, invented the first photovoltaic cell by placing silver chloride in an acidic solution and connecting it to platinum **electrodes**. When this device was exposed to light, it generated an electric current producing the photovoltaic effect, also known as the Becquerel effect. This is the basic process through which a photovoltaic cell converts sunlight into electricity. Although Becquerel could not explain this phenomenon, his work provided the basis for solar cells. Charles Fritts, in 1883, constructed the first working example of a solar cell. The device used a metal plate and thin layers of selenium and gold. Although the design functioned, its efficiency was less than 1 percent.

Wave Theory

In 1873, James Clerk Maxwell published *Treatise on Electricity and Magnetism*. Maxwell had studied (and published) the works of Henry Cavendish, a reclusive eighteenth-century English physicist and chemist. Cavendish had done experiments that determined the relative conductivity of a series of metals. Further incorporating the work of Ampere, Øersted, and especially Faraday, Maxwell used mathematics to provide a unifying theory of electrical and magnetic behavior. Maxwell's equations, as they are called, proved the existence of electromagnetic waves and demonstrated that the speed of these waves must be identical to the speed of light. He realized that light had to be an electromagnetic wave. Eight years after Maxwell died, Heinrich Hertz conducted an experiment that demonstrated the existence of electromagnetic waves, just as Maxwell had predicted.

Maxwell's equations, complex and difficult to understand, are still considered to be one of the most elegant and important scientific achievements. As Richard Feynman, one of the twentieth century's most well-known physicists, said:

> The great transformations of ideas come very infrequently … we might think of Newton's discovery of the laws of mechanics and gravitation, Maxwell's theory of electricity and magnetism, Einstein's theory of relativity, and … the theory of quantum mechanics.

Magnetic Fields and Diodes

In 1879, Edwin Hall invented a device to measure the strength of a magnetic field. Hall's revolutionary discovery was not applied for several decades, but when it was finally recognized, it provided an essential technique for a wide variety of innovations. These range from position and motion detectors to automobile crankshaft sensors to space rocket engines. The Hall effect is also essential for characterizing the conductivity of semiconductors.

In 1874, Ferdinand Braun observed that, at the point of contact between a metal point and a galena crystal, current flows freely in only one direction. This was the first description of a semiconductor **diode**, which is a device that allows only **direct current** flow. This discovery, a few decades later, contributed to the invention of the radio.

Arthur Schuster, a German-born British physicist, also noticed the process of **rectification** in a circuit created with copper wires that had been left unused for a period of time. When Schuster cleaned off the copper wires (removing the copper oxide that had formed while the circuit was unused) the rectification effect was gone. Schuster had identified copper oxide as a new semiconductor.

The Hall effect sensor

Electrons!

J. J. Thomson, in 1897, discovered electrons and opened new possibilities in understanding atomic structure and its relationship to conduction. A few years later, in 1900, Paul Drude, a German physicist, developed what is known as the Drude model. The Drude model describes electrical conductivity in metals and the movement of electrons. Drude incorporated Maxwell's work in electrodynamics in his theory. His model attempted to explain the distinction between a metal and an insulator, the transport of electricity in metals, and thermal conductivity in metals. According to Drude, every metal contains a large number of free electrons that float through a lattice of positive **ions.** The electrons move in random directions, colliding with the lattice ions. The basic difference between conductors and nonconductors is that conductors have a greater number of free electrons. Drude explained that the resistivity in metals increases when the temperature increases because the electrons become more scattered. This explained the behavior of metals, but it did not explain what happened in other materials where the conductivity increased when the temperature increased.

The model provided only for short-term random interactions between electrons and their environment. Eduard Reicke further developed the Drude model, and his theory included both negative and positive charge carriers with different mobilities and varied concentrations. The flaws in this model were not addressed until the development of quantum mechanics provided a new theory for explaining the behavior of electrons.

USING ELECTRICITY: From MOTORS to VACUUM TUBES

While scientists were investigating the nature of electricity, engineers and inventors were finding uses for the phenomenon. As theoretical understanding was expanding, commercial applications were providing important and world-changing uses. As they increased knowledge, they also provided an incentive for further research.

The Telegraph

After Ørsted identified the magnetic action of electrical currents, there were numerous inventions based on his discovery. In the United States, Samuel Morse, an artist who loved tinkering with electric devices, worked on creating a machine that sent electrical signals along a wire that connected stations: the telegraph. Morse also invented a machine that used a pencil attracted by an electromagnet to place dots and dashes (representing letters of the alphabet) on a moving tape: Morse code. These inventions made it possible to instantly send complex messages along the telegraph lines. Morse sent his first message, from Washington, DC, to Baltimore, Maryland, in 1844. It said, "What hath God wrought?" By 1866, a telegraph line had been laid on the floor of the Atlantic Ocean, connecting America to Europe. Although now almost obsolete, the telegraph laid the groundwork for future communications innovations.

Willoughby Smith, a London engineer, had the responsibility for laying several underwater telegraph cables, including one under the English Channel, as well as the first one in the Mediterranean Sea. He then helped lay the transatlantic cable between Ireland and Newfoundland. Smith investigated using selenium to insulate the cable, and the results of his experiments were reported in *Nature,* February 20, 1873:

> The early experiments did not place the selenium in a very favorable light for the purpose required, for although the resistance was all that could be desired … yet there was a great discrepancy in the tests, and seldom did different operators obtain the same result … While investigating the cause of such great differences

Laying the transatlantic cable

> in the resistance of the bars, it was found that the resistance altered materially according to the intensity of light to which they were subjected ... Merely intercepting the light by passing the hand before an ordinary gas-burner ... increased the resistance from 15 to 20 percent. If the light be intercepted by glass of various colors, the resistance varies according to the amount of light passing through the glass.

Smith's experiments enabled him to devise a way to use selenium, a semiconductor, as an effective underwater insulator. The experiments also led to using selenium as a photosensitive material in a variety of devices.

The Telephone and Photophone

In 1876, Alexander Graham Bell was granted a patent for the telephone, an instrument that used an electric current to transmit the human voice through wires. Bell, though, thought the photophone, based on transmitting sound through a beam of light and first used in 1880, was his most important invention.

Bell used selenium in the first photophone (later known as a radiophone). While the telephone was developing acceptance, the photophone had more difficulty being universally adopted because of a variety of technical limitations. It was, however, the first wireless communication device, and it predated the radio by almost twenty years. Bell's pioneering invention was also a forerunner in the use of **fiber optics**, which, beginning in the 1970s, transformed the communications industry.

Radios

Maxwell's equations, in the 1860s, had provided a theoretical indication of the existence of radio waves and their ability to move through free space. Maxwell, however, never conducted tests to illustrate his proofs.

In 1887, Heinrich Hertz, a German physicist, described the first experiment that proved the existence of electromagnetic waves. Hertz said:

> From the start, Maxwell's theory was the most elegant of all ... the fundamental hypothesis of Maxwell's theory contradicted the usual views, and was not supported by evidence from decisive experiments.

Hertz began by conducting a series of experiments with sparks, noting that they produced a charge in the wires they jumped between. These charges, if Maxwell was correct, should radiate electromagnetic waves that would pass through the air. Hertz constructed a device that created oscillating electric charges and, through electromagnetic waves, produced electric current within a copper wire located 4.9 feet (1.5 meters) away.

Hertz had proven Maxwell's thesis. He had passed electrical energy through the air, without wires. In the following years, Hertz conducted further experiments that demonstrated that the waves from his oscillator had the same properties as light waves, indicating that radio waves and light waves were both part of what we now call the electromagnetic spectrum.

Although Hertz was right about electromagnetic waves, he was wrong about their importance. He did not believe that his discovery had much practical application. But his work was the basis for many of the technological innovations, including radio, television, and mobile phones, that ushered in the modern age of communications. Hertz's work stimulated a number of scientists to explore its uses.

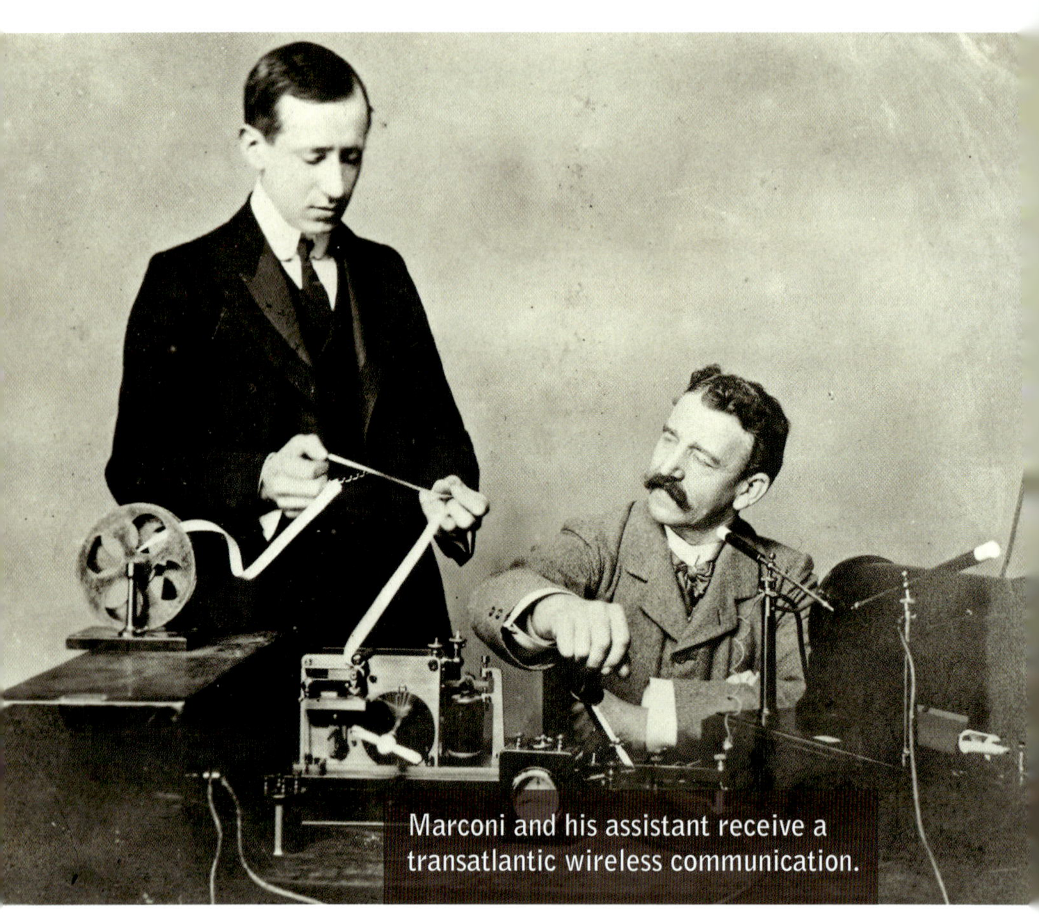

Marconi and his assistant receive a transatlantic wireless communication.

In 1894, Jagadish Chandra Bose, an Indian physicist, used exploding gunpowder to ring a distant bell. His demonstration proved that electromagnetic waves can communicate without using wires. Bose did not patent his work, and he did not attempt to use it commercially. Bose was also the first to use crystals to detect radio waves and patented the first semiconductor device for that use.

Guglielmo Marconi, an Italian inventor, took out the first patent in 1896, in England, for what was known as radio telegraphy. But a Serbian American, Nikola Tesla, had already been working on a theoretical model for a radio. In 1943, the US Supreme Court reviewed a 1915 decision based on a lawsuit by Tesla and awarded him Marconi's original radio patent, even though Tesla had never built a working radio. Tesla had, however, built the Tesla coil, a device for sending and receiving radio waves. In 1895, Tesla's lab burned down before he was able to complete his scheduled radio transmission.

Meanwhile, based on Hertz's work, Marconi had created a radio wave device at his family's estate in Italy. The Italian government was not impressed by Marconi's invention, so he decided to see if he could stir up more interest in London. Within a year of his arrival, Marconi had financial backers and was broadcasting for a distance of 12 miles (19.3 km). In 1885, Thomas Edison took out a patent on a system of wireless communication between ships, based on induction, not radio waves. Marconi's company bought the patent, despite its different technology, in order to prevent any other challenges to Marconi's work.

By 1899, Marconi's company was transmitting across the English Channel. Next, Marconi worked on sending transatlantic radio messages. Scientists had thought that radio waves traveled in straight lines and could not be broadcast beyond the horizon. Marconi thought, incorrectly, that they followed Earth's curvature. The wave path, however, did resemble a curve because the waves bounced off the ionosphere and returned toward Earth. Although Marconi's understanding was wrong, nonetheless he was able to make the first transatlantic transmission, from Cornwall, England, to St. John's, Newfoundland, in 1901. Marconi's company demonstrated the usefulness of radio communication for military and shipping applications.

Cat's Whisker

Early radio receiving devices did not have to transmit information by sound. They used what was known as radio telegraphy: spelling out messages in Morse code, with dots and dashes being communicated by different pulse lengths of radio waves. The crystal detector device was one of the most successful in early radio. One of the first of these detectors was known as the cat's whisker (sometimes called the crystal detector).

The cat's whisker was developed by several radio researchers, including Henry Dunwoody, using a silicon crystal, and G. W. Pickard, using silicon carbide. The device used a thin wire resting on a crystal of semiconductor material (usually galena) creating a basic **point-contact rectifier**. This functioned to change the **alternating current** being transmitted into direct current. The cat's whisker was one of the first semiconductor diodes—an electronic component with two electrodes that allows current to move in one direction—and was, in fact one of the first semiconductor electronic devices. When AM radio transmission was developed, it was discovered that crystal detectors could receive audio transmissions as well as telegraphy.

By the 1920s, the crystal detectors (which continue to have some uses today) were largely replaced by amplifying receivers using the vacuum tube. Although they were more sensitive and more powerful and required less continuous tunings, vacuum tubes were more expensive and needed powerful batteries to

A cat's whisker was one of the earliest radio transmission devices.

run them. Crystal detectors were used by the military and commercial markets for almost ten more years.

Vacuum Tubes

Vacuum tubes became the basic component of most electronic devices in the first half of the twentieth century. In 1904, John Ambrose Fleming demonstrated a device called the Fleming diode, which converted alternating current into direct current. Fleming's invention was based on an observation, "the Edison effect," that Thomas Edison had made but did not put to practical use.

This early vacuum tube was based on the incandescent lightbulb, which has a filament sealed in an airtight container usually made of glass. The Fleming diode has an extra electrode inside. When heated, the filament emits electrons into the vacuum. The second electrode, or plate, if it has a more positive voltage, will attract these electrons. This flow is only in the direction from filament to plate because since the plate is not heated, it does not emit electrons. The filament and plate together create an electric field.

This diode, a type of vacuum tube with only two electrodes, is used in the process called rectification. This conversion from alternating current (AC) to direct current (DC) occurs when the current can pass in only one direction. It has many applications, and one of the earliest was its use in the demodulation of AM radio signals, a process that enhances transmission.

Other inventors tried unsuccessfully to improve on Fleming's device. In 1907, though, Lee de Forest patented a vacuum tube based on the Fleming diode but with an additional electrode. This grid consisted of a bent wire between the filament and the plate. When de Forest applied the wireless telegraph signal to the grid rather than the filament, he was able to detect signals with greater sensitivity. This process altered (modulated) the current and served as an electronic

LIGHT-EMITTING DIODE

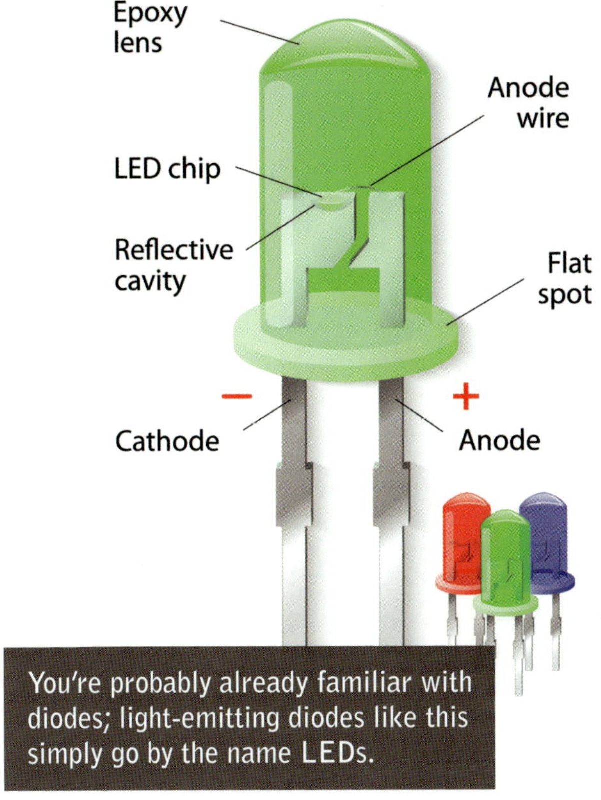

You're probably already familiar with diodes; light-emitting diodes like this simply go by the name LEDs.

amplifier. The Audion, as de Forest's device was known, made the transmission of sound possible (rather than the transmission of signals) and made radio broadcasting live.

Between de Forest's creation and the 1960s, an enormous number of vacuum tubes, most based on de Forest's work, were developed, both for specific and more general functions. Although **solid-state electronics** are the norm today, vacuum tubes still have certain specific uses. Because transistors are more efficient only at low frequencies, some high-power radio stations continue to use vacuum tubes. Then, too, there are consumer markets for vacuum tubes in certain items, such as guitar amplifiers. But, for the most part, the vacuum tube has been replaced by the solid-state component, and the ever-expanding electronics boom no longer includes the vacuum tube.

Alessandro Volta was the first to use the term "semiconductor."

CHAPTER 2

The Science of Semiconductors

The understanding of semiconductors began as pure scientific inquiry, with scientists noting peculiarities in the conductivity experiments they were undertaking. Eventually the theory became intertwined with discoveries about the ways to use semiconductors and the astonishing things they could make possible. Theories led to some practical applications and, in turn, other inventions led to more theoretical understanding. It was a gradually unfolding process, until it exploded, in 1947, with the invention of the transistor and the dawn of the Information Age.

While we cannot identify a moment when the science of semiconductors became totally clear, we can instead identify the moment when the actual use of semiconductors led to a communication revolution. And we can understand how that moment happened by examining the history of semiconductor research up to that point.

SEMICONDUCTOR PROPERTIES: EARLY OBSERVATIONS

Alessandro Volta, an Italian physicist, used the term "semiconductor" for the first time in 1782 when he touched a charged electrometer to a variety of different materials. His results indicated that different materials produced either an immediate discharge (conductors), no charge (insulators), or a charge within a short period of time (semiconductors). In 1833, an English scientist, Michael Faraday, was the first to observe the effect of heat on a semiconductor. Faraday knew that in metals, such as copper, electrical conductivity decreased with an increase in temperature. But, when Faraday applied heat to silver sulfide, its conductivity increased. The result was just the opposite of what he had expected. His papers on these experiments provide the first recorded examination of the semiconductor effect. Scientists later came to realize that increasing the temperature of most semiconductors increases the density of charge carriers within the material, and this increases their conductivity. But, for Faraday, this was an extraordinary and unusual observation that he could not fully explain.

Soon thereafter, scientists began studying the effects of light on conductivity in certain materials. In 1839, Edmond Becquerel observed that shining light on one of the electrodes in an electrolytic cell affected its conductivity and produced a **photovoltage**. Then, in 1873, Willoughby Smith described the phenomenon of "photoconductivity" that resulted from illuminating a sample of selenium. The sensitivity of selenium to light was tested in many experiments—it was observed that selenium was even sensitive to moonlight. In 1876, William Adams and Richard Day observed the light effect on a dry semiconductor, as opposed to an electrolyte solution, when they experimented with the contact on a selenium sample.

Carl Ferdinand Braun, a German physicist, investigated the conductivity of the point of contact between metal and crystals of certain semiconductor substances, such as lead sulfide (galena) and iron sulfide (pyrite). He discovered that the joint reduced the voltage flowing in one direction and allowed greater current flow in the other direction. This was the first example of a rectifier that was able to change alternating current to direct current.

All these observations enabled nineteenth-century scientists to identify some of the peculiar phenomena associated with semiconductors. They noted that as temperatures are increased, the conductivity of a semiconductor increases. They also found that illumination of some semiconductors produces a photovoltage and increases conductivity. And, finally, they saw that contact with certain semiconductors results in rectification.

All these observations increased interest in semiconductors. They also made it clear that certain aspects of semiconductors and their behaviors required more attention and investigation. Although much was becoming known, scientists did not know how to effectively measure the characteristics of semiconductors, and they still did not have a theoretical understanding of how semiconductors actually worked.

HALL EFFECT

Edwin Hall began his career as a high school principal in Brunswick, Maine. He then enrolled in a doctoral program in physics at Johns Hopkins University and, while preparing for his doctoral thesis, he discovered, in 1879, what is known as the Hall effect. With a sensor he created, Hall was able to measure the invisible area of force around a magnet. Hall began his experiments using gold leaf on a glass plate. When he introduced a magnetic field to a current traveling across the metal strip, he discovered that a voltage difference was produced across the metal strip. This Hall effect results when the perpendicular magnetic field exerts a force on the particles in the electric current and displaces the charges at right angles

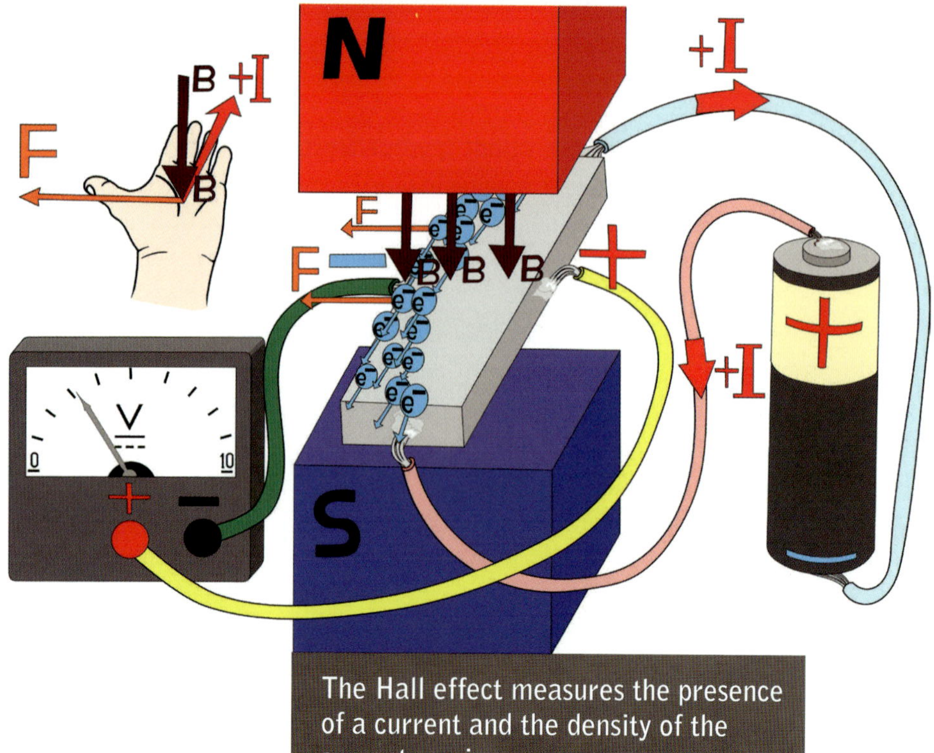

The Hall effect measures the presence of a current and the density of the current carriers.

to both the magnetic field and the flow of current. The charge deflects in a curved path as it moves through the substance because the charge's own magnetic field interacts with the magnet's field.

Hall then conducted similar experiments on semiconductors as well as metals. Hall discovered that the voltage developed is directly proportional to the current, the magnetic field, and the characteristics of the conducting material. The Hall effect can measure both the presence of a current in a magnetic field and the density of the current carriers. Different materials have different Hall coefficients, a mathematical statement that is specific to the material of the conductor. These different materials develop different Hall voltages, even though they are the same size and are exposed to the same magnetic field and the same electric current. In semiconductors, the sign of the Hall voltage indicates whether a positive or negative charge predominates.

DISCOVERY of the ELECTRON

At the time of Hall's work, scientists did not really understand how an electric charge was carried. By the end of the nineteenth century, J. J. Thomson, a British physicist, discovered the explanation. Thomson was born in Manchester, England, on December 18, 1856, to a third-generation antiquarian bookseller and the daughter of a cotton manufacturer. Thomson was a shy boy who displayed an early scientific interest. At the young age of fourteen, he was admitted to Owens College, known for its outstanding scientific studies. His parents had planned to apprentice him as an engineer with a locomotive company, but when Thomson's father died, the family was unable to fund the apprenticeship. Then at age nineteen, Thomson obtained a mathematics scholarship to Cambridge University.

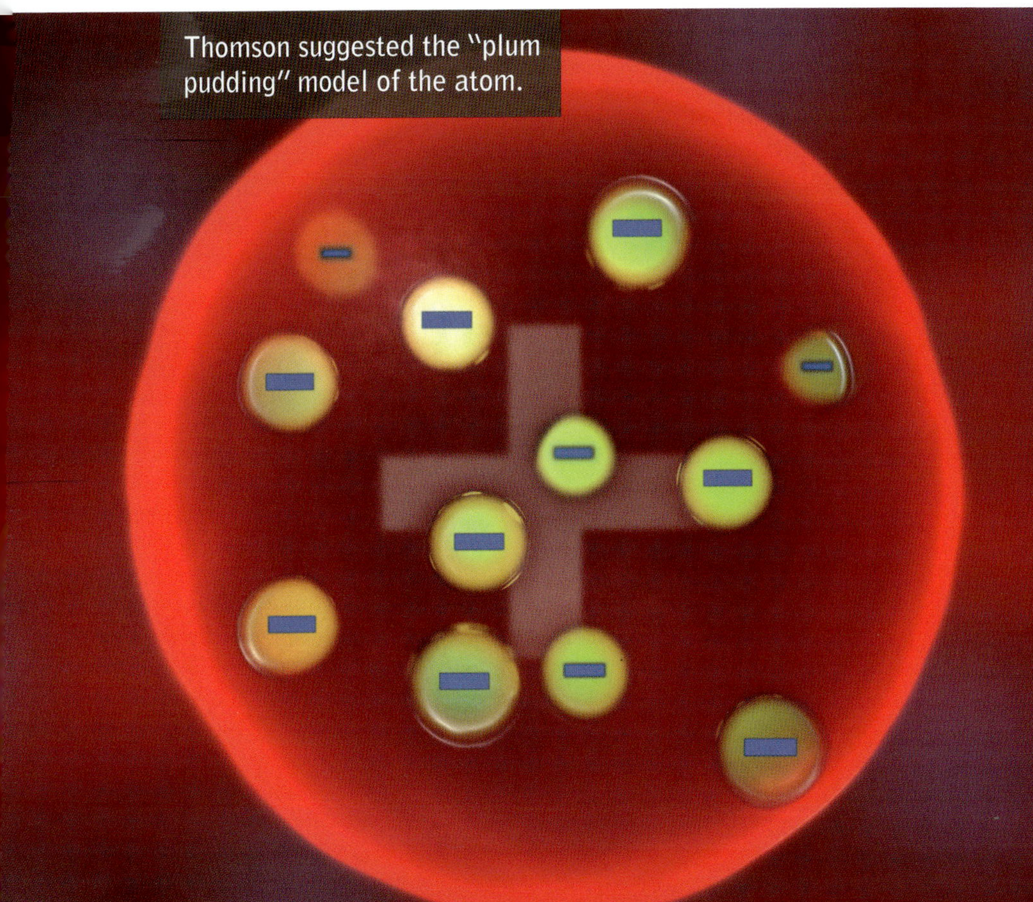
Thomson suggested the "plum pudding" model of the atom.

After receiving his master's, Thomson became a fellow of Trinity College, Cambridge, in 1881. In 1882, he won the prestigious Adam's Prize in mathematics and then, somewhat surprisingly, was selected over older and more experienced laboratory scientists to become Cavendish Professor of Physics. Thomson soon proved, with his extraordinary talent, that he deserved the recognition.

His work on electrons provided a critical element in the understanding of electric conduction and a building block on which all later understanding was based. This discovery was not only essential to the theory and the use of semiconductors, it was essential to understanding the structure of all matter.

After Michael Faraday coined the word "ion" to describe particles that were attracted to charged electrodes, scientists understood that atoms were associated with electric charges. But atoms were still thought of as indivisible particles. Thomson was the first to suggest that one of the basic units of matter was not the atom, but a particle more than one thousand times smaller than an atom. He explored this thesis through his experiments with **cathode rays**.

Although Thomson was considered clumsy, he was expert at designing experimental apparatuses. In 1897, in his investigations of the properties of cathode rays, he applied an improved vacuum technique in his cathode ray device. Scientists knew that by changing the voltage in a cathode ray tube they could emit rays that produced an image on a phosphorescent surface. They did not, however, understand how this worked. Some speculated that it was the result of waves traveling from the **cathode** through the vacuum in the tube. Others decided that the rays were particle streams. Thomson investigated by placing cathode ray tubes into magnetic fields. He understood that the magnetic fields had little effect on how a wave moves, although they could move particles. In his experiments, the cathode rays moved, bending over to one side. He realized that the ray must be composed of small particles.

Thomson then decided to investigate the nature of these particles. The particles were too small for him to be able to directly make calculations, but he could approximate their size from seeing how much they were bent when electrical currents of different strengths were applied. Thomson realized that the particles either carried an enormous charge or that they were a thousand times smaller than the particle then known to be the smallest atom: hydrogen. The particles could travel farther through air than an atom-sized particle could. He also measured the heat generated when the ray contacted a thermal junction. He decided that these rays were actually composed of particles that came from within the atom itself, and he called those particles "corpuscles." This was a bold new idea.

Thomson, for the first time in history, realized that atoms could be divided. He suggested a model of the atom, known as the "plum pudding" model. He thought of the atom as a sphere of positive matter and hypothesized that electrostatic forces controlled the positions of the corpuscles within it. In Thomson's model, the corpuscles—soon to be known as electrons—were rapidly orbiting within this sphere. As Thomson said:

> At first there were very few who believed in the existence of these bodies smaller than atoms. I was even told long afterwards by a distinguished physicist who had been present at my lecture at the Royal Institution that he thought I had been pulling their legs.

Thomson wasn't joking. He was the first to identify the electron and discover subatomic particles.

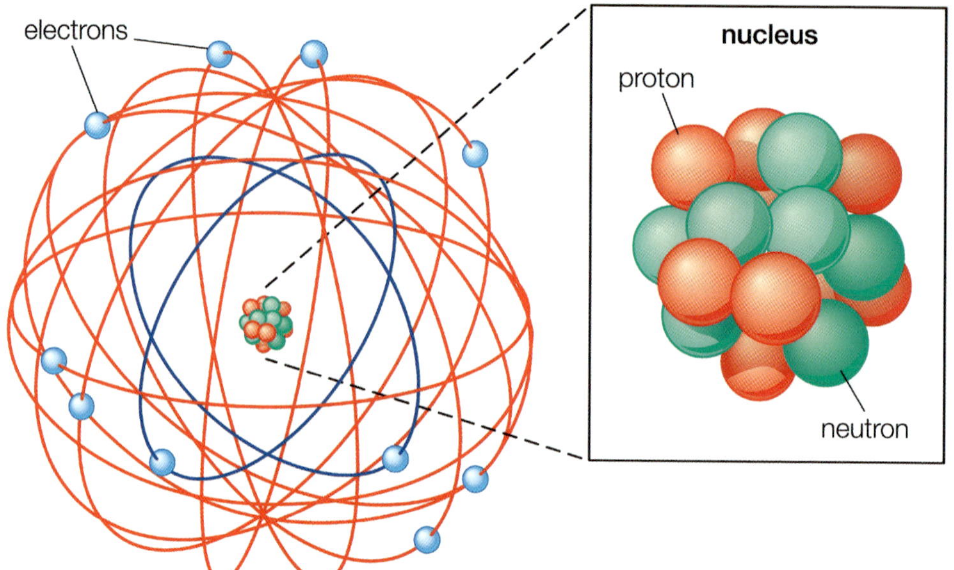

Rutherford's planetary model of the atom described it as a miniature solar system because electrons orbit the nucleus.

RUTHERFORD'S MODEL of the ATOM

Although Thomson's discovery was monumental, other scientists did not accept all the aspects of his explanations. In fact, Thomson's plum pudding theory of the structure of the atom was challenged by a student of Thomson's, Ernest Rutherford.

Rutherford was born and educated in New Zealand. After receiving his B.A. from Canterbury College in New Zealand, he won a scholarship to attend Cambridge University. He arrived in England in 1895 where he began his studies with Thomson. Rutherford had already built an advanced radio receiver when he was in New Zealand, but in England he decided to focus on work he found more intellectually stimulating and began studying radiation. He used the terms "alpha," "beta," and "gamma rays" to describe the most common radiation types.

His radiation studies resulted in him being called the father of nuclear physics. After teaching physics in Canada and in Manchester, England, Rutherford returned to Cambridge in 1919. Upon Thomson's retirement, Rutherford succeeded him as Cavendish Professor of Experimental Physics at the University of Cambridge.

While Rutherford was still at Manchester, he supervised two of his assistants, Hans Geiger and Ernest Marsden, in carrying out a famous and groundbreaking experiment. The "gold foil" experiment, as it was known, involved barraging metal foil with alpha particles and observing how they scattered. The investigators used a sample of radium to provide the alpha particle stream. Based on Thomson's atomic model, they expected the particles to pass through the foil without being deflected, since the model said that the diffuse electric fields in the atom would not significantly affect the alpha particles. However, the researchers discovered that particles were being deflected by more than 90°.

Rutherford explained this by suggesting that the positive charge of the atom is concentrated in a tiny nucleus at its center and that the nucleus has a very heavy mass. He thought that the atom might be like a miniature solar system, with a larger positively charged center surrounded by a few corpuscles—now called electrons. The electrons orbited at a distance from the core of the atom and were much lighter than the core. Later, the core nucleus was found to consist of two kinds of particles: protons and neutrons.

Rutherford's planetary model of the atom was more accurate than Thomson's and accounted for some characteristics that could not be explained in the plum pudding model. But Rutherford's model, too, did not explain certain critical features of atomic behavior. Rutherford's model was unable to account for the atom's stability. The electrons moving around the nucleus of the atom emit radiation as they move, and subsequently they lose energy. As they lose energy, they should slow down and gradually spiral toward the nucleus. At that point, the atom would collapse. But, despite the appeal of Rutherford's model, it was clear that the atomic structure is stable and the atoms do not collapse. A new explanation of atomic structure was still necessary. That would come from **quantum theory**.

SEMICONDUCTOR CHARACTERISTICS

Meanwhile, in Germany, Johann Königsberger was working on questions of conductivity. Königsberger examined the problem of substances whose conductivity increases when the temperature is increased.

In 1906, Königsberger initially wrote that:

> When the temperature of oxides and sulphides is raised, the number of electrons, i.e. of free conducting quanta of electricity is greatly increased until it reaches its limit. Then their behavior reminds that of the metals in which at normal temperature, almost all the electrons are free.

In 1914, Königsberger expanded the picture of semiconduction and described a class of semiconductors, substances with more in common than just being neither good nor poor conductors, but somewhere in between. These materials, according to Königsberger, have a temperature-conductivity dependence that could be explained by processes occurring within the crystals: a magnitude of resistivity within certain limits; much higher values of thermos-emf (voltage developed by heat) than seen in metals; and a significant sensitivity to light. All these characteristics identified a whole new class of substances and provided the modern classification

The Life of James Clerk Maxwell

Although James Clerk Maxwell is not well known outside scientific circles, he is considered to be one of the most influential thinkers in the history of science. His work altered the basic perception of physics and laid the foundation for its modern age. Albert Einstein called Maxwell's work "the most profound and fruitful that physics has experienced since the time of Newton."

Maxwell was born in Edinburgh, Scotland, in 1831, an only child of parents who had married late in life. When Maxwell was an infant, the family moved to a country house that his father had inherited, and Maxwell began his education with a boring and lackluster tutor who thought that Maxwell was not very smart. But, after his mother's death, Maxwell was sent to the Edinburgh Academy.

Maxwell was a curious student and had an astonishing memory. He published his first scientific paper at age fourteen, demonstrating his great interest in geometry. He enrolled at the University of Edinburgh at age sixteen and then, in 1850, he went to Cambridge University.

Although he was involved in research in optics and gas, Maxwell is best known for his groundbreaking theories in electromagnetism, demonstrated in the famed Maxwell's equations. These elegant mathematical statements paved the way for the theory of relativity and quantum mechanics and greatly influenced modern physics. Maxwell died on November 5, 1879, at the same age and of the same abdominal cancer that had killed his mother.

of semiconductors. It offered new pathways for scientific exploration and applications.

QUANTUM MECHANICS

In 1900, Max Planck, a German physicist, originated the quantum theory, which demonstrated that light was emitted or absorbed in discrete amounts, or "quanta," contradicting the accepted idea that light was a continuous electromagnetic wave. Quanta of light became known as photons, and the idea that these photons could have the dual properties of waves and particles was examined.

Louis-Victor de Broglie and Electron Waves

In 1924, Louis-Victor de Broglie, a French physicist from a noble family, became a scientist despite a family tradition in the diplomatic service. When he heard about the theories of Plank and other physicists, de Broglie refused to continue with a research project in French history. Instead he focused his energies on understanding conceptual problems in physics. In 1924, de Broglie proposed a revolutionary idea—that atomic matter might have the property of waves.

Twenty years before, Einstein had suggested that short wavelength light could behave as if it were made up of particles. De Broglie extended the idea that light had dual properties to suggesting that ordinary matter also had dual properties. He observed that, for most particles of matter, the wavelength might not be detectable but that electrons should have a measurable length.

Experiments had shown that there were restrictions in the motion of an electron around a nucleus, but the reasons for those restrictions were not understood. De Broglie's theory, that

German physicist Max Planck is the originator of quantum theory.

the electron had the property of a wave, provided a possible explanation for that restriction. If a wave was limited by atomic boundaries, its shape and its motion would be restricted. Although there were no experiments verifying de Broglie's insight at the time, it was a major accomplishment in theoretical physics.

Einstein, who got a copy of de Broglie's thesis, was impressed and used it in continuing his own work. Erwin Schrödinger, an Austrian physicist, then constructed a mathematical system of wave mechanics. In 1927, George Thomson, a Scottish physicist, and Clinton Davisson and Lester Germer, American scientists, conducted experiments that provided the first evidence of the wave property of electrons. Quantum mechanics was proving to have the power of predicting atomic behavior.

Neils Bohr and Werner Karl Heisenberg

Neils Bohr, a Danish physicist, began his career with practical experiments on surface tension. Soon, however, his interests became much more theoretical. Bohr went to England and worked with two atomic structure pioneers, J. J. Thomson and Ernest Rutherford. In 1913, based on Rutherford's ideas about the nucleus of the atom, Bohr began developing his own atomic model. Bohr's description of atomic structure incorporated new concepts, based on quantum theory. In Bohr's model, the electrons orbited the nucleus but the range of possible orbits was limited. The electrons were only able to move between these orbits under specific conditions.

This provided an explanation for the atom's stability where Rutherford's model did not. Before Bohr, the atom's nucleus was described as a heavy core with a positive charge, and negatively charged electrons orbited around this nucleus, following a planetary path. Basing his idea on the structure of hydrogen, Bohr adjusted the planetary view of the motion of electrons to accommodate the pattern of light emitted by hydrogen atoms.

Niels Bohr devised a model of atomic structure based on quantum physics.

Bohr hypothesized that, if the electrons were limited in their orbital paths, the discrete wavelengths of light the atom emitted could be explained. This light, then, radiated from the hydrogen atom when an electron jumped from an outer orbit to an orbit closer to the atom's nucleus. The energy it lost in this jump is equal to the energy in the quantum of emitted light.

Under a Rockefeller grant, from 1924 until 1925, Bohr worked at the University of Copenhagen with Werner Heisenberg. Heisenberg was a German pioneer in the field of quantum physics and became noted for his famous uncertainty principle. This stated that the more you know about the position of a particle, the less you know about its motion, and the more you know about its motion, the less you know about its position. Influenced by Heisenberg, among others, Bohr continued to include quantum physics concepts into his model. Bohr's picture of atomic structure still serves as a limited but useful explanation of both the chemical and physical characteristics of elements. While it explained some characteristics of the hydrogen atom, however, it was not completely accurate in describing other elements.

The Electron Cloud

As quantum mechanics developed and additional experiments were devised, it became more clear that describing electrons and their orbits in terms of particles did not provide a complete picture. It was essential to consider electrons as having properties of waves as well as particles. This understanding resulted in a new conceptualization of electrons and their activities—the electron cloud.

In this model, electrons do not follow a planetary orbit but exist as stationary waves— mathematically expressed as the sum of two waves being created from opposite directions. Because of this wave-like property, electrons are never completely at a single point. As a result of particle-like properties, however, each wave state has the same electrical charge as the electron particle. And electrons jump between orbits as though they were particles.

The planetary analogy of atomic structure, as though it is the same as planets revolving around the sun, has certain limits because electrons have properties of particles *and* waves. A more useful picture is provided by thinking of the electron as a large and oddly shaped cloud around the atom's nucleus. The elliptical planetary path of the electron, described by Bohr, only applies in an atom that has a single electron. If more electrons are added, they will fill in the space more evenly and create the electron cloud in a generally spherical zone. This model is based on the uncertainty principle and is built on probability rather than knowledge that can be verified.

The STRUCTURE of SEMICONDUCTORS

Empirical inquiry had fueled early efforts to understand the behavior of semiconductors. Theoretical concepts were then formulated on the structure of atoms. The introduction of quantum mechanics into the study of semiconductors provided some explanation for the behavior of electrons in both metals and semiconductors. One of the most influential of these quantum theorists was Alan Wilson.

Alan Wilson and Band Theory

The band theory of semiconductors was developed by Alan Wilson in 1931, incorporating the principles of quantum mechanics. Wilson was the son of a maintenance engineer for a ferry company in the north of England. After gaining a scholarship to attend his local grammar school, Wilson prepared for a commercial career by taking evening courses in bookkeeping and shorthand. By the end of World War I, however, changes in higher education made it possible for

Wilson to continue his studies. He received a scholarship and was accepted at Emmanuel College, Cambridge, where he undertook a degree in mathematics.

At Cambridge, Wilson learned about Heisenberg's uncertainty principle and the theories of quantum physics. He became a research fellow and produced several papers related to quantum mechanics. Wilson had joined the Kapitza Club, an invitation only group of mathematicians, where he learned of the European perspective on theoretical physics. He received a Rockefeller Fellowship that enabled him to travel and work with both Heisenberg and Bohr. During this trip, Wilson formulated his groundbreaking theory of semiconductors and developed an explanation for why semiconductors increased their conductivity when the temperature was increased.

Wilson realized that conceptualizing electrons as existing in energy bands could account for substances being classified as either conductors, insulators, or semiconductors. Wilson saw these energy bands as having energy gaps between them. He described a valence band, an area completely occupied by electrons, and the conduction band, an empty band beneath the valence band. In Wilson's proposition, energy was needed to move electrons from the valence band to the conduction band. This supported Königsberger's idea that energy activation was responsible for conduction in semiconductors.

In discussing the relationship of heat to conduction, Wilson stated that, when the temperature is raised:

> There will be a few electrons in the second band and conduction can take place … The increase in the number of conduction electrons with rise of temperature will tend to increase conductivity, while the excitation of the thermal vibrations of the lattice will tend to decrease it. At low temperatures the first effect must predominate, and so resistance will have a negative temperature coefficient. At higher

temperatures the effect of the thermal vibrations will be the more important and resistance will obey the normal law. This is just what is observed for semiconductors.

Inconsistencies in behavior could be explained in terms of vacancies in the valence bands. This led to the concept of "**holes**"—unoccupied spaces—in the valence band, an idea that Heisenberg had also introduced. Wilson additionally discussed the idea of impurities in relationship to semiconductors and calculated various properties of the solids he was studying.

Wilson continued his work on solids and wrote two papers on rectification in metal semiconductor junctions and in electrolytic rectifiers. In the *Theory of Metals*, which Wilson published in 1936, he elaborated on what was known, at that time, about metals, insulators, and semiconductors.

Wilson's model offered an account of the role of impurities and demonstrated how they could improve conductivity. In fact, according to Wilson, "the observed conductivity of semiconductors must be due to the presence of impurities." He stated that impurities whose electrons had energy states that were closer to the energy state in the conduction band than to the state in the pure substance would increase conductivity. This understanding provided the basis for the process of **doping**, in which impurities were deliberately added to semiconductors to enhance conductivity.

Wilson's theory of the band structure of electrons and his ideas about the function of impurities were essential concepts that provided the foundation for all the progress in the use of semiconductors that was to follow. The transistor, the integrated circuit, and the microchip could not exist without Wilson's groundbreaking understanding.

Semiconductors being manufactured at Fairchild Semiconductor's bustling assembly line in 1970

CHAPTER 3

The Major Players in the Discovery

The use of semiconductors created a technological revolution that is still unfolding. Many scientists, including Nobel Prize winners, contributed to the understanding and practical knowledge that made this revolution possible. And although this technological explosion got its great momentum after World War II, scientists began doing experiments and wondering about the nature of semiconductors early in the nineteenth century.

MICHAEL FARADAY and the FIRST RECORDED SEMICONDUCTOR EFFECT

In the autumn of 1791, Michael Faraday was born in Newington Butts, England, the third of four children of humble parents. He was raised in the Sandemanian sect of the Christian Church and received little formal education. He became, though, one of the most respected scientists of his age. His experiments and analyses provided the basis of modern electro-technology.

Michael Faraday was the first to record the semiconductor effect in his experiments on silver sulfide.

At the age of fourteen, he was apprenticed to George Riebau, a London bookbinder. Using every opportunity to read many of the books in Riebau's shop, Faraday immersed himself in scientific studies and attended lectures at the Royal Institution, an organization founded to introduce new technologies and teach the public about science. When he completed his bookbinding apprenticeship, Faraday wrote to Sir Humphry Davy, the first professor of chemistry at the Royal Institution.

Davy was initially reluctant to hire him. When Davy's eyesight was damaged in a chemical accident, he did take Faraday along as an assistant on a special trip to Europe, where Davy met with other famous scientists. Faraday, not considered a gentleman by the social standards of the time, was treated as a servant and was so unhappy that he considered going home and giving up his scientific aspirations. When they returned to London, however, Faraday was appointed as chemical assistant at the Royal Institution. Faraday then rose steadily, appointed as superintendent of the house in 1821 and director of the laboratory in 1825.

Faraday's scientific enquiries began with chemical analyses, and he investigated many substances, including chlorine, benzene, steel alloys, and optical glass. He provided the first description of the optical properties of **nanoparticles**. His most famous work, however, concerns electricity and magnetism. Faraday succeeded, in 1821, in building a device that produced what he called "electromagnetic rotation" that became the foundation of the electric motor. When Faraday did not give credit to his mentor in the development of this mechanism (something that Davy had attempted unsuccessfully to build), their relationship was damaged. For a time afterward, Davy limited Faraday's involvement in electromagnetic experiments, although Faraday continued his correspondence on the subject with scientists he had met when he earlier traveled in Europe with Davy.

After Davy's death, Faraday's investigations of electromagnetism expanded. He began a series of experiments in which he discovered electromagnetic induction, a way of producing an electrical current in a circuit using the force of a magnetic field. He explored the process of electrolysis, which passes electrical current through a substance in order to create a chemical change. He soon completed a series of experiments that led him to revise the understanding of the nature of electricity. Faraday proposed that electricity was a single phenomenon and not, as many scientists at the time believed, a set of distinct phenomenon. He also developed the theory that electricity was the result of particles responding to a variety of magnetic forces. He challenged the then-current ideas of the nature of matter and established the field theory of electromagnetism, which eventually became a foundation of modern physics.

Faraday was also a skilled teacher and lecturer. He began a series of Christmas Lectures for children and Friday Evening Discourses. Both series (which have continued to the present day) focused on providing scientific education to the public. He was able to communicate complicated ideas in clear and simple terms. As a devout Christian, he also spoke frequently both as a deacon and elder in the Sandemanian church where he met and married his wife, Sarah Barnard. Several biographers have noted that Faraday's faith had a strong influence on both his life and his work.

In 1833, during one of Faraday's experiments, he made the first documented observation of what is now known as a semiconductor. During an investigation of the effect of temperature on silver sulfide, Faraday noted that the conductivity of the silver sulfide increased as the temperature was increased. This response is typical of semiconductors. It is the opposite of the effect noted in metals, like copper, where a temperature increase results in a decrease in conductivity.

Faraday's extraordinary career was surprising considering his basic formal education. It included a steady output of groundbreaking experiments and theories, and membership in the American Academy of Arts and Sciences, the Royal Swedish Academy of Sciences, the French Academy of Sciences, and the Royal Netherlands Academy of Arts and Sciences. He twice refused to become president of the Royal Institution. The farad, the unit of measuring electrical capacitance, is named after him. He is considered by many as one of the most influential scientists in history. Confirming his place as an instrumental force in modern science, Faraday's picture hung on the wall of Albert Einstein's study.

WILLIAM SHOCKLEY, WALTER BRATTAIN, JOHN BARDEEN, and the INVENTION of the TRANSISTOR

William Shockley was a controversial figure who was both honored for his scientific achievements and despised for his beliefs on racial superiority. Shockley was a critical figure in the development of the semiconductor. During his career, he was awarded many honors, including the Nobel Prize. Yet he died estranged from most of his colleagues and family members.

Shockley was born on February 13, 1910, in London, England. His father was an MIT-trained mining engineer whose family descended from *Mayflower* pilgrims, and his mother was a Stanford graduate and the first woman surveyor in Nevada. After a financial failure in London, the family moved back to the United States and settled in Palo Alto, California. There, Shockley's indulgent parents educated him at home until he was eight.

William Shockley developed the junction transistor.

From the Faraday Files

In 1821, Faraday wrote a letter to August De La Rive, a Swiss physicist, that accompanied his electromagnetic rotation device. Faraday spoke of what he called "a little apparatus" and described how it worked:

> Royal Institution November 16, 1821
>
> Dear Sir,
>
> Herewith you will receive copies of my papers which I mentioned in a letter I sent to you, per post, a month or two ago, and which I hope you will do me the favor to accept. I also send in the packet a little apparatus I have made to illustrate the rotatory motion on a small scale. The rod below is soft iron, and consequently can have its inner end made north or south at pleasure by contact of the external end with one of the poles of a magnet. To make the apparatus act, it is to be held upright with the iron pin downwards; the north or south pole of a magnet to be placed in contact with the external end of the iron pin, and the wires of a voltaic combination connected, one with the upper platinum wire, the other with the lower pin or magnet; the wire within will then rotate, if the apparatus is in order, in which state I hope it will reach you. Good contacts are required in these experiments.
>
> ... I am extremely busy, too much so at present to try my hand at anything more, or even to continue this letter many lines further; but I hope soon to have a little news on steel to send you.
>
> I am, my dear Sir, as ever, your very obliged and faithful,
>
> M. Faraday

In 1928, Shockley enrolled in the California Institute of Technology, majoring in physics. Pursuing his PhD, he entered MIT in 1933, where he soon became known for his scientific talents. While there, he also met and married his first wife, Jean Alberta Bailey. Philip Morse, Shockley's mentor at MIT, arranged for him to work at the famous Bell Laboratories, the research and scientific development center originally founded by Alexander Graham Bell. His work there involved a variety of experimental inquiries including research on energy bands in solids, **photoelectrons** in silver chloride, and the theory of vacuum tubes.

When World War II began, Shockley was again recruited by his mentor Morse and went to work as research director of the Antisubmarine Warfare Operations Research Group. There his work helped increase the success of Allied attacks on the German U-boats that had been controlling the oceans. He worked on training bomber crews for the Army Air Force and helped increase their effectiveness. Shockley became one of the highest-ranking civilian scientists and received the National Medal of Merit. But the personality problems that were to follow him throughout his life eventually resulted in his alienation from his friends and colleagues. His marriage was falling apart, and he even threatened to kill himself. His work was his focus.

When the war ended, he returned to his work at Bell Labs. Shockley assembled a team with Walter Brattain, already working at the center, and John Bardeen, recruited by Shockley. They focused on attempting to find a replacement for the vacuum tube. Although they had the same goal, Brattain and Bardeen worked together developing a point-contact transistor. Shockley went off on his own and shortly afterward developed the **junction transistor**. Although it was not the first transistor, Shockley's device was easier to mass-produce and use in varied applications.

Bell Labs gave credit for the invention of the transistor to the entire team, although Shockley's name was not on the original patent. The subsequent confusion about who should get credit contributed to the increasing discord between Shockley and the other scientists. Brattain refused to work for Shockley again, and Bardeen quit Bell Labs. But the Nobel Prize for the invention of the transistor went to all three men.

In 1953, Shockley left his wife, Jean, and he also left Bell Labs and took a job at Caltech. Soon thereafter, financed by entrepreneur Arnold Beckman, Shockley founded Shockley Semiconductor Laboratory to manufacture silicon-based semiconductor transistors. He located the company in California's Santa Clara Valley, near Stanford University. Shockley's firm was the beginning of the technological revolution that came to be known as Silicon Valley.

Shockley hired a group of talented and intense scientists to work at Shockley Semiconductor. But less than a year after he won the Nobel Prize, Shockley's teamed mutinied, infuriated by Shockley's arrogant and disrespectful attitude. The group left Shockley to form Fairchild Semiconductors. Some of them, a few years later, founded Intel. All of them became quite wealthy. Shockley did not. Shockley remarried and accepted a teaching position at Stanford University in 1958. From 1963 to 1965, he served as the Alexander M. Poniatoff professor of electrical engineering and applied sciences. But conflict and controversy continued to follow him. Shockley began promoting his theory of a "retrogressive evolution." Although he lacked any training in genetics, he insisted that intelligence was genetically determined and claimed that black people were intellectually inferior. He even proposed financial rewards to encourage the "genetically disadvantaged" not to have children.

Shockley's outrageous ideas and hateful comments on race overshadowed his scientific accomplishments. Yet he was a major contributor to the technological revolution that resulted in computers, cell phones, and other innovations of the modern age.

Walter Brattain

Walter Brattain was the first of the three men credited with designing the transistor to come to work for Bell Labs. Brattain was born in Amoy, China, where his father had a job teaching. The family moved back to the state of Washington, and Brattain was raised there on a cattle ranch. Brattain's time on the ranch gave him practical skills, and he was noted for being good with building things.

After his time studying physics at Whitman College, the University of Oregon, and the University of Minnesota, Brattain went to work for the National Bureau of Standards as a radio engineer. Brattain was interested in doing work in his field of physics. At an American Physical Society meeting he was introduced to Joseph Becker from Bell Labs. Brattain left to join Bell Laboratories as a member of the technical staff.

Brattain began his work at Bell Labs studying the surface properties of metals, and his early work was on tungsten. He was assigned to look at photo effects on semiconductor surfaces and rectification at the surface of copper oxide. He did similar studies on silicon.

John Bardeen

John Bardeen was born in Madison, Wisconsin, in 1908. His father was an anatomy professor at the University of Wisconsin Medical School. Bardeen attended University High School but graduated from Madison Central High School at the age of fifteen, despite difficulties caused by the early death of his mother. He began studying electrical engineering at the University of Wisconsin and graduated in 1928 with a BS, having taken a term off to work in the engineering department at Western Electric.

Bardeen continued his studies as a research assistant for the next two years, working on applied geophysics and antenna radiation. He first studied quantum theory during this time with Professor J. H. VanVleck. Bardeen followed his geophysics teacher, Leo Peters, to Gulf Research Laboratories where they worked on methods for applying magnetic and gravitational surveys to searching for oil.

Bardeen, though, realized that he was more interested in theory than in applied science, and he enrolled in a mathematical physics program at Princeton University. He began his study of solid-state physics and got his doctorate. He then went to Harvard as a junior fellow and worked again with Professor VanVleck. Their work was on electrical conduction in metals and the density of nuclei.

During World War II Bardeen worked at the Naval Ordnance Laboratory, and after the war, he joined Shockley's solid-state research group at Bell Labs.

ROBERT NORTON NOYCE and the INVENTION of the INTEGRATED CIRCUIT

Robert Norton Noyce was born in Burlington, Iowa, on December 12, 1927, and was brought up in Grinnell, Iowa. His father, Ralph, was a clergyman, and his mother, Harriet, was the daughter of a clergyman. Both of Robert's parents graduated from Oberlin College and were noted for their intelligence.

The third of four sons, Robert demonstrated an early aptitude for science. At age twelve, he and his brother built a

Robert Noyce was a cofounder of Intel.

small aircraft. Later, he attached an old engine to the back of his sled and built a homemade radio. He attended Grinnell College, where he was an outstanding diver on the swim team, played the oboe, and acted in dramatic productions. In his junior year, he was almost expelled from school for stealing the mayor's pig and roasting it at a school luau. His physics professor, unwilling to have a brilliant student expelled, negotiated a deal in which Robert paid for the pig and got suspended for a semester.

After a six-month suspension, Noyce returned to Grinnell. Grant Gale, Noyce's physics professor, discussed a new invention from Bell Labs: the transistor. In their joint exploration of the reports on this device, Noyce learned, for the first time, about how the transistor utilized the qualities of semiconductors. Noyce then graduated Phi Beta Kappa with a BA in mathematics and physics.

He entered MIT, enrolling in the physics doctoral program, and received his degree in 1953. After graduation, Noyce worked briefly at Philco Electronics. Inspired by the idea of working with Shockley, an inventor of the transistor, Noyce took his wife and children to California, bought a house, and met with Shockley to ask for a job. Noyce got the job, but the relationship, given Shockley's arrogant behavior, was strained from the start.

Noyce wasn't the only employee at Shockley Semiconductor who found working with Shockley difficult. A group of seven scientists decided to leave Shockley and form their own organization. They recruited Noyce, a natural leader, and became known (according to Shockley) as the "Traitorous Eight." Although the group had been struggling to obtain funding for their new company, when Noyce made a passionate presentation to the CEO of Fairchild Industries, Sherman Fairchild decided to bankroll the new corporation, Fairchild Semiconductor. This began a Silicon Valley pattern of technology entrepreneurs spinning off new companies at a speed that equaled the speed of new technological developments.

At Fairchild, Noyce was the general manager. His working style created a new atmosphere for employees—casual, supportive, and open to innovation—that became the signature style for Silicon Valley companies. While as an executive Noyce created a new corporate culture, he continued, as a scientist, to work on improvements to the transistor.

Based on a process devised by Jean Hoerni, another member of the Traitorous Eight, Noyce invented a method of manufacturing an entire circuit—composed of transistors, **resistors**, and **capacitors**—on a single semiconductor wafer.

In 1968, discouraged by the resistance of Fairchild Semiconductor's parent company to organizational changes that would give employees even more control and more rewards, Noyce left Fairchild, along with Gordon Moore and Andrew Grove, and founded Intel. The new company planned to make semiconductor chips for memory storage. BUSICOM, a Japanese calculator company, asked Intel to make a calculator chip. But Ted Hoff, at Intel, envisioned a more complex and comprehensive product. After only nine months of development and design, overseen by Noyce, Intel announced the world's first microprocessor—a chip that incorporated storage and the ability to program logic and arithmetic operations. Almost every modern electronic device, from cell phones to automobiles to washing machines, uses the microprocessor. Intel became the leading microprocessor producer in the world.

In 1974, Noyce was divorced from his first wife, Elizabeth, with whom he had four children. That same year, he married Ann Schmeltz Bowers. Bowers was the first director of personnel for the Intel Corporation and subsequently the first vice president of human resources for Apple, Inc.

Noyce continued with Intel, first as president, then as chairman of the board of directors. In 1978, he left Intel to become chairman of the Semiconductor Industry Association. In 1988, he became the first president of Sematech, a consortium to support American semiconductor manufacturing.

Noyce held many semiconductor patents, and his accomplishments were recognized with many awards. Among other honors, he was awarded the IEEE Medal of Honor and the National Medal of Science, inducted into the US Business Hall of Fame, and elected fellow of the American Academy of Arts and Sciences. He was known as the "mayor of Silicon Valley" because of his many contributions to the technological and cultural innovations that characterized that dynamic locale.

A former swimming champion and avid athlete, Noyce died of a heart attack in 1990, after his daily swim.

JEAN HOERNI and the PLANAR PROCESS

Jean Hoerni, the physicist responsible for the process that put Fairchild Semiconductor's integrated circuits at the forefront of the new technologies, was born in Geneva, Switzerland, on September 26, 1924. The son of a Swiss banking family, Hoerni earned a BS in mathematics and a PhD in physics from the University of Geneva, and another PhD from Cambridge. In 1952, he came to California as a postdoctoral research fellow in chemistry at the California Institute of Technology.

Hoerni's early work was theoretical and dealt with such things as the quantum theory of solids, but he was lured away from his academic career to join William Shockley's firm, Shockley Semiconductor. After only a year, Hoerni, along with seven other Shockley scientists, formed Fairchild Semiconductors.

While at Shockley, Hoerni's office was separate from the other scientists, who were focused on improving transistor technology. On his own, Hoerni worked on theoretical calculations. At Fairchild, he continued his solitary analyses.

While the other Fairchild founders were organizing the new company, Hoerni was devising a revolutionary approach to manufacturing transistor circuits.

Fairchild had begun manufacturing a commercial product called mesa transistors, so named because they had layers resembling the rock layers of the mesas in the American Southwest. The drawback of this design was that the places where the layers interfaced, and where the electrical activity occurred, were exposed at the edges. Hoerni suggested a solution that was completely contrary to the manufacturing process then in use. Hoerni's **planar process** suggested that the exposed junctions at the surface of the silicon transistors could be protected by "oxide masking techniques" that would prevent contamination and leakage.

Although the Fairchild team did not have the expertise to implement the planar process when Hoerni first introduced it in 1957, by early 1959 they had fabricated a prototype planar transistor. It resolved the problems of the unsuccessful mesa transistors and provided superior stability and performance. Fairchild centered its transistor production on the planar process, and it became the basis of the integrated circuit. Disillusioned with the policies of Fairchild's parent company, in 1961 Hoerni left Fairchild. With another member of the Traitorous Eight, Jay Last, he started another semiconductor company, the Amelco division of Teledyne. When his position at Teledyne was reduced, in a cost-cutting maneuver, from general manager to director of research, Hoerni looked for other alternatives and went to work for Union Carbide. Then, in 1967, he founded Intersil, making innovative low-power, low-voltage microchips for digital watches.

While remaining involved in the semiconductor industry as an investor and consultant, Hoerni became more involved in charitable activities. His lifelong interest in mountaineering connected him to philanthropic organizations, such as the Central Asia Institute, which he cofounded in order to support community education in impoverished Asian villages. Hoerni died in 1997, survived by his wife, Anne Marie, and their three children.

Although the planar process might not be as famous as the transistor or the integrated circuit, Hoerni's invention has been called "the most important innovation in the history of the semiconductor industry."

Diodes and transistors are critical advances in semiconductor technology.

CHAPTER 4

The Discovery of Semiconductors

The story of semiconductors is not really the story of a single discovery or one moment when a single great idea led from ignorance to understanding. It is, instead, the story of a series of linked experimental discoveries and progressive theoretical advances. Yet, as John Orton says, in the *History of Semiconductors*:

> If one had to choose a single event which truly put semiconductors on the international map, it would surely be the invention of the transistor at Bell Telephone Laboratories in late 1947. Without this, the leap into "information technology," which has so radically changed all our lives, may never have occurred.

When the transistor was invented, it sparked greater interest in semiconductors. Electronic research until that point had focused on vacuum tubes. Semiconductors had played a minor role, except in the World War II research on detectors for radar systems. Once the transistor became a reality, however, semiconductor research took off. It has not slowed down yet.

The INVENTION of the TRANSISTOR

Attempts were made to design a transistor as early as 1926. Various semiconductor substances were used and various designs attempted. There were even patents for devices that never were created. For the most part, though, serious scientific and engineering efforts concerted on improving vacuum tubes.

Bell Laboratories

Bell Telephone Laboratories, the scientific center founded by Alexander Graham Bell, was the research and engineering arm of the American Telephone and Telegraph Company. Its main work was to plan and support the telephone system's design and components. The practical aspects of the work at Bell Labs, however, was based on an extensive and solid research focus. It brought together specialists from a variety of professions and theoretical experts in a variety of fields. Three of those specialists were Walter Brattain, William Shockley, and John Bardeen. They jointly were awarded the Nobel Prize for inventing the transistor. The work at Bell Labs revolutionized communications.

Shockley's Team

Mervin Kelly, the director of research at the labs, had begun a program of hiring outstanding young physicists to staff the research projects. As early as 1936, when he first

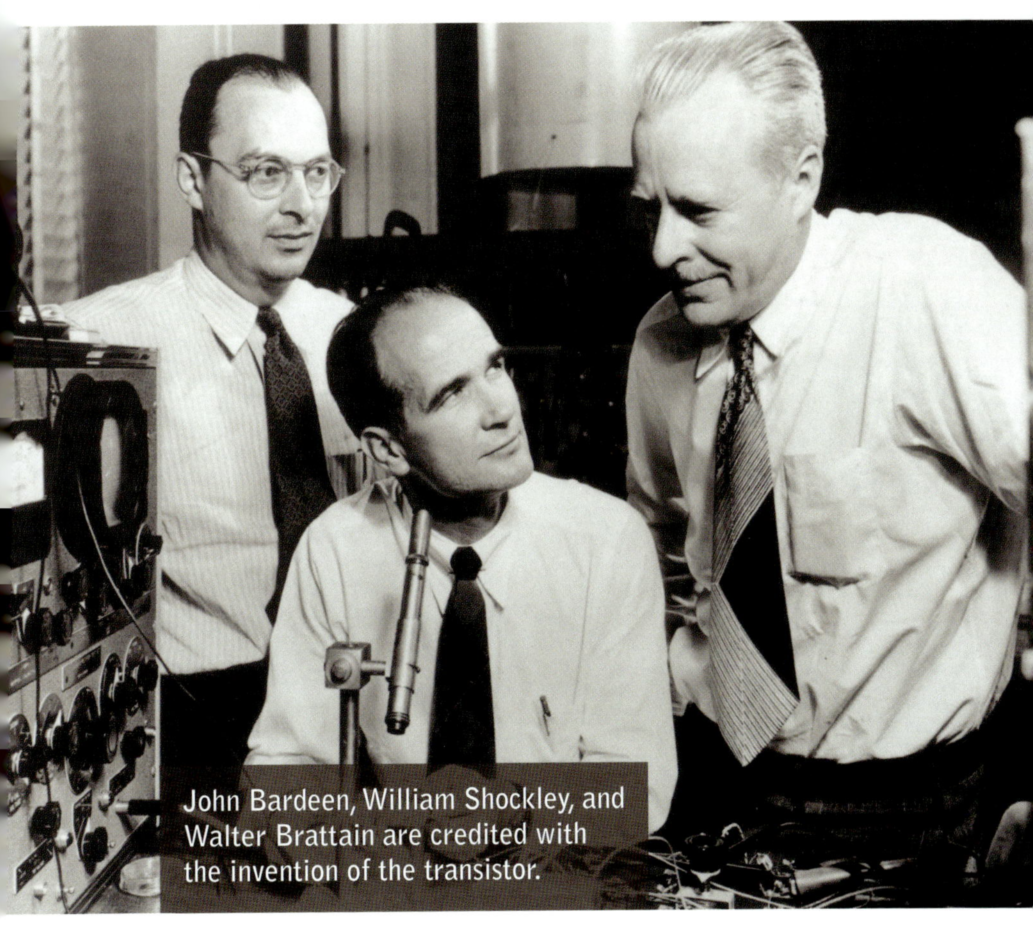

John Bardeen, William Shockley, and Walter Brattain are credited with the invention of the transistor.

hired Shockley, Kelly had announced that he was not just interested in new vacuum tube technology but in completely new electronics. More and more, the scientists were turning to focusing on the behavior of electrons in solids and studying semiconductors. When Shockley returned to Bell Labs after World War II, he began assembling a scientific team to find a replacement for the vacuum tube. Shockley's new team consisted of Walter Brattain, already at Bell Labs, and John Bardeen, recruited by Shockley. Like Brattain, Shockley began his work at Bell with vacuum tubes, but he also got involved in research on band theory and photoelectrons and silver chloride. Bardeen had been studying solid-state physics and worked at Harvard on electrical conduction in metals.

The P-N Junction

Brattain was one of the few scientists at Bell who appreciated the work of Russell Ohl. Ohl started working at Bell Labs in 1927. At that time, radio tuners could pick up only the lowest frequencies, and Ohl wanted to create one that could handle higher frequencies. He thought semiconductors might be the answer. Ohl's work was understood by only a handful of scientists in the organization, and the Bell Lab management tried to shift his focus several times. Yet Ohl persisted, and he investigated various semiconductor materials, such as silicon and **germanium**.

In 1939, Ohl began experimenting with an early electronic component, the cat's whisker, also called a crystal detector. Ohl tried to eliminate the unreliability of the device by growing purer versions of silicon crystals. But when he improved the purity of the crystals, he discovered that they no longer had the ability to function as a detector. One day he noted that one of his purest crystal samples actually worked well in the cat's whisker. This sample, though, had a crack down the middle.

Ohl noticed that as he moved around trying to test it, the device would work and then stop working for no apparent reason. He realized that the exposure to light was altering the conductance in the crystal and increasing the current flow between the two sides of the crack. Ohl showed the cracked sample to several colleagues, including Walter Brattain, who immediately realized that the crack formed a sort of junction.

Ohl continued to work on resolving the mystery and deduced that the crack was a dividing line that had occurred when the melted silicon was hardening in the purification process. Different impurities in the silicon had been accidently isolated into different regions, with the crack separating them. According to Ohl experts at ETHW:

> As a result, the silicon atoms in the region on one side of the crack had extra electrons around them. The other region was the opposite; its crystallized silicon had a slight shortage of electrons. They named the two regions p and n—p for positive-type and n for negative-type. The barrier between the impurities was called the p-n junction. The junction represented a barrier, preventing the excess electrons in the n-region from traveling over to the p-region, where atomic forces naturally drew them.

Although this idea eventually became the basis of the major developments in semiconductor technology, it was not immediately pursued.

The Invention

Shockley's team had one goal: replace the vacuum tube. Although they had the same goal, Brattain and Bardeen worked together developing a point-contact transistor; Shockley worked alone to create the junction transistor. Brattain and Bardeen made a terrific team. Brattain, with his ranch-hand practicality would design and implement their experiments. He acknowledged that he "had an intuitive feel for what you could do in semiconductors, not a theoretical understanding." Bardeen, with his eye on the theoretical implications, would explain what had happened and why it had happened.

Bardeen knew that the accepted theories of conductivity in semiconductors stated that both the polarity and the magnitude of the current that traveled through a semiconductor were the same in all parts of that substance. He suggested, instead, that the electrons on the surface of the semiconductor behaved differently than the other electrons in the solid. Bardeen proposed that electrons were immobilized at the surface of the semiconductor. He also said applying an electric field did not alter the conductivity within the whole of the solid. The scientists realized that these surface properties governed the process of converting an alternating current to a direct current (rectification).

Now that they were equipped with an understanding of surface effects, Bardeen and Brattain worked with germanium to conduct experiments to change its surface potential. While investigating the surface states, they found that if they placed a small positive charge on one of two closely spaced electrodes, they would greatly increase the surface conductivity. If they could control what was occurring on the surface of the germanium, they could build an amplifier that worked.

Germanium and silicon are elemental semiconductors, and they have orderly atomic structure. They are less complex than selenium or certain other compounds such as copper oxide. Scientists at Bell Labs expected to more easily be able to process materials that showed the least crystal complexity. The scientists had chosen to work with germanium because it was easier to modify than silicon. It could be manipulated at lower temperatures, and it was less subject to contamination.

The device they constructed consisted of two closely spaced metal points carefully piercing a germanium slab. Since available wires were too thick for the purpose, Brattain placed a strip of gold foil on a plastic triangle over the germanium. With a single slice, he created two minimally separated contacts. One of the points was the emitter, and the other was the collector. A third contact, the base, was attached to the germanium slab and was the voltage source. When current was applied to one of the contacts, a stronger current came out from the other.

On December 23, 1947, their efforts finally succeeded. They had built the first transistor.

Shockley's response was not wholehearted jubilation. He wrote:

> The birth of the point-contact transistor was a magnificent Christmas present for the group as a whole. I shared the rejoicing. But my emotions were somewhat conflicted. My elation with the group's success was tempered by not being one of the inventors. I experienced some frustration that my personal efforts, started more than eight years before had not resulted in a significant inventive contribution of my own.

Shockley was distressed about not having participated in the invention, and when Bell applied for the patent, Brattain and Bardeen were named and Shockley was not. But Shockley's contribution was acknowledged, and all the publicity about the invention included Shockley as one of the three team members.

The Junction Transistor

Shockley, though, continued to believe that there could be a better transistor. He wanted to produce a device that was less fragile than the point-contact transistor, easier to manufacture, and with a higher power capacity. Before the first of the new year, Shockley began working on a new transistor design. He came up with the new design very quickly.

Shockley described his transistor as composed of "at least three layers having different impurity contents." Shockley's design avoided the unreliable wires that made point-contact transistors hard to regulate. Yet, in order to have electricity flow through the semiconductor crystal, the germanium had to be very pure.

Gordon Teal, a scientist at Bell but not a member of Shockley's team, was working on the chemical-electrical properties of germanium and silicon. Teal thought that transistors should use a single semiconductor crystal, rather than a slab cut from a larger crystal. Shockley was not initially interested in Teal's idea. When he saw that current flowing over Teal's crystal lasted over one hundred times longer than it did over the semiconductor slabs, he realized Teal was right.

The large crystals needed for this transistor were formed using a painstaking method that involved dipping a tiny seed crystal in melted germanium. The action of a junction transistor is produced by creating a sandwich of **n-type semiconductors** (semiconductors in which the conductivity is due primarily to the movement of electrons) and **p-type semiconductors** (semiconductors in which the conductivity is due primarily to the movement of positive holes). The layers in Shockley's design were achieved by adding certain impurities to the melt and first forming a p-type layer. They then quickly added a different impurity, resulting in an n-type layer.

A patent for Shockley's design was filed in 1948, but it took another year before the first junction transistor was manufactured and another two years before a reliable device was produced. By the middle of 1951, several usable junction transistors had been made. Although it was not the first transistor, Shockley's was the most versatile and the easiest to mass-produce.

Bell Labs had given credit, but not the original patent for the point-contact transistor, to the whole team of Brattain, Bardeen, and Shockley. And Shockley got the patent for the junction transistor, but his other team members did not.

Promoting and Developing the Transistor

Although the scientists at Bell Labs knew they had a remarkable invention, there was a lot of work that had to be done before the transistor could demonstrate its usefulness and even before its full impact could be understood. Bell's early publicity for the device exceeded its immediate technological possibilities. The transistor was promoted as a replacement for vacuum tubes in both radio and television, but its ability to function was initially quite limited. For a number of years, the transistor was essentially a laboratory curiosity.

In the early 1950s, both the point-contact transistor and the junction transistor were manufactured. Yet the advantages of the junction transistor soon became apparent and the point-contact transistor was phased out. Bell gave licenses to a number of firms to manufacture transistors, and other

Electrons, Holes, and Conductivity

Electrons and holes are present in any semiconductor substance. Each atom has a fixed number of negatively charged electrons that occupy orbits (bands) based on their energy. The outermost band that contains electrons with the highest energy is called the valence band. Its electrons are called valence electrons. The empty band outside the valence band is called the conduction band. The conduction band is the orbit from which electrons can escape from, or be accepted by, the atom. The space between these two bands is called the band gap.

A hole is not a physical particle like an electron. Instead, it is the absence of an electron in a particular place in the atom, and it has a positive charge. Like electrons, holes can be passed from place to place in an atom—from one band to another—and electrons and holes can both be passed between atoms.

A hole forms when an electron moves from the valence band to the conduction band. This hole can be filled by another electron, which itself leaves a hole in the location it moved from. This process, as electrons and holes move around within the semiconductor material, contributes to conductivity.

A process known as doping—adding minute amounts of impurities to a semiconductor—increases the number of charge carriers in the material. Some impurities increase the number of electrons, and others increase the number of holes. The resulting negatively or positively charged materials are essential in the production of solid-state electronics.

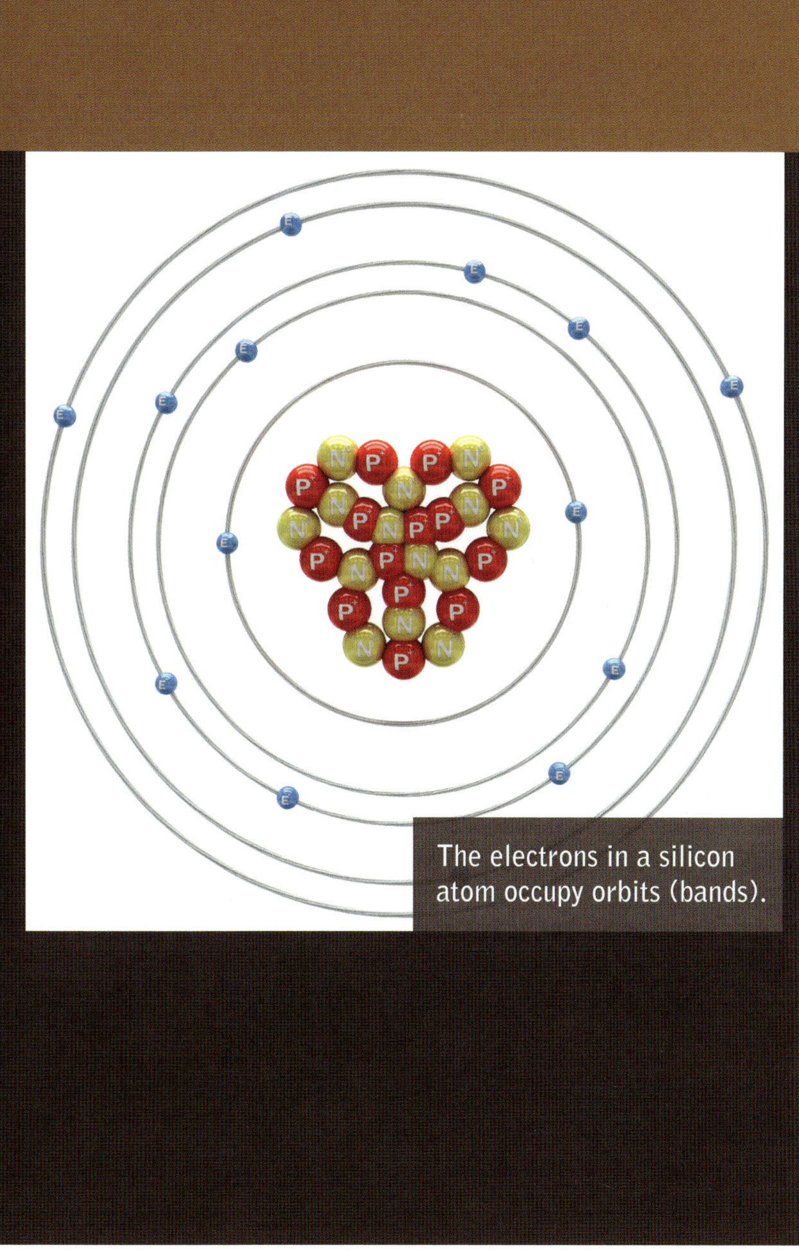

The electrons in a silicon atom occupy orbits (bands).

companies began to manufacture them without any license or training. At many different sites, the transistor got the benefit of additional research and a wider technological experience.

The first large-scale use of the junction transistor was in hearing aids. For these tiny devices, the transistor was a clear winner over the vacuum tube because it was much smaller, used less power, and was more efficient and more responsive. But the problems of producing reliable and functional transistors were apparent. Early transistors were noisier than vacuum tubes, more restricted in their range, and more liable to damage. Then, too, it was difficult to manufacture any two transistors with the same characteristics, and there was little quality control. Most transistor manufacturing was provided by the same organizations that produced vacuum tubes—there may have been resistance to adopting a new technology that threatened to replace a familiar one.

Even though the transistor was in commercial production, it required the continuing attention of scientists to assure its success and continued development. In the early 1950s, transistors typically used geranium as the semiconductor substance. Germanium and silicon were both relatively simple structures to understand in terms of quantum theory, so most of the research on semiconductors concentrated on these two substances.

The researchers initially concentrated on germanium because it was easier to manage, could be worked at lower temperatures, and was not as susceptible to contamination. But its manufacturing was expensive and the transistors were somewhat fragile. When the size of semiconductors became more important, however, the characteristics of silicon made it the preferred material.

In 1954, Texas Instruments announced that it had made a silicon transistor. Silicon could work at much higher temperatures than germanium and eventually replaced germanium in most semiconductor applications.

Other improvements in transistor manufacture continued throughout the decade. Processes for growing semiconductor crystals, regulating the additions of impurities, and masking the semiconductor to control placement of the impurities added to the progress in creating reliable transistors.

In 1951, there were four US companies manufacturing transistors. By 1956, there were more than twenty-six. Then William Shockley decided to leave Bell Labs and move his transistor development work to California. He chose to start his new laboratories in a peaceful area of Santa Clara County where he had grown up.

Shockley Semiconductor

When Shockley decided to found his own company for semiconductor research and development and the manufacture of advanced components, he recruited a diverse group of chemists, physicists, and engineers. As Gordon Moore, one of the scientists at Shockley Semiconductor, described:

> It was pretty much research on developing the basic technology, trying to repeat some of the things that had been done [at the Bell Labs], understand some of the problems that we still had, because neither the processing nor the physics of [silicon] was well understood. We were just exploring the technology and figuring out what could be done and we had a lot of things to make work before we could try to build something. We were far from developing a commercial device.

Soon, though, what became the Silicon Valley pattern of organizational mobility and technological innovation was evidenced in the formation of Fairchild Semiconductors by several members of Shockley's team, headed by Robert Noyce. There, the next signature discovery in the story of semiconductors—the integrated circuit—was developed. Although the invention of the transistor is the most critical moment in the history of semiconductors, the invention of the integrated circuit comes in a close second place.

The INTEGRATED CIRCUIT

All electronic equipment was taking the path of miniaturization. Transistors were getting smaller and smaller. However, the ways in which those transistors had to be connected—the wires and other components needed to do the job—limited just how small the electronic units could be. Building an effective circuit required that all the connections be intact or the current flow would fail. In the late 1950s, transistorized circuits were constructed manually, with each small component being soldered in place and connected by metal wires. Assembling the enormous numbers of tiny components needed for a device like a computer would be increasingly difficult to achieve without an increasing likelihood of connection failures. Then as the circuitry became more complex, its increasing size increased the distance the electric signals had to travel and slowed the response time of the components.

In 1958, in Dallas, Jack Kilby, a newly hired electrical engineer at Texas Instruments, began working on a new concept. Kilby set out to make all the elements of a circuit from semiconductor materials and build an entire circuit, not just the transistor, from a single crystal. He soon had built a model, which looked like a piece of germanium glued to a glass slide with wires sticking out of it. The model worked, though, and Kilby filed a patent for the new integrated circuit in February 1959.

At Fairchild Semiconductors, at about the same time, Robert Noyce was also working on constructing an entire circuit on a single semiconductor chip. Based on Hoerni's planar process, Noyce devised a way to create an entire circuit—transistors, resistors, capacitors, and their connectors—on a small piece of silicon. He modified the planar design by placing the aluminum interconnections on top of the oxide layer, insulating them from the silicon below them.

While Kilby awaited the results of his patent application, Noyce submitted his own extremely detailed application, hoping that it would not conflict with Kilby's. Noyce's application was actually given the first patent, but both men are now acknowledged to have thought of the same revolutionary idea at the same time. Noyce's integrated circuit (IC), though, proved to be more effective and more commercially successful. When Kilby received the Nobel Prize in 2000 as coinventor of the integrated circuit, he acknowledged Noyce's achievement. Since the awards are not given posthumously, Noyce was no longer eligible.

The integrated circuit is responsible for both the ever-increasing capacity of electronic devices and their decreasing cost. In the first decade of integrated circuit production, the transistor count per integrated circuit increased from ten to four thousand. In the next decade, the number of transistors in an integrated circuit went from four thousand to five hundred thousand. By the third decade, the count was at one hundred million.

While the transistor count on each integrated circuit increased by ten-million-fold, the cost of creating each chip had hardly increased at all. Thus, the invention of the integrated

The integrated circuit is composed of connected transistors.

circuit made complex electronic devices available not just to large industrial purchasers but to individual consumers.

The INFORMATION REVOLUTION

In 1968, both Noyce and Gordon Moore—another member of the Traitorous Eight that left Shockley to found Fairchild Semiconductor—had become unhappy with the policies of their parent company. Fairchild Camera and Instrument Company had provided start-up funding for Fairchild Semiconductor and now exercised managerial control. Noyce had attempted to expand organizational changes that would increase employee control and rewards at the semiconductor company. Fairchild Camera and Instruments resisted the changes. Noyce and Moore left Fairchild together, along with Andrew Grove, another Fairchild researcher. The three scientists founded Intel.

Andrew Grove was a survivor of the Nazi Holocaust and the 1956 Soviet invasion of his native Hungary. Handicapped by a severe hearing loss, he was destitute when he arrived in the United States. Nonetheless, he got his BA from the City College of New York and then his doctorate from the University of California at Berkeley. In 1997, he was chosen Man of the Year by *Time* magazine for being "the person most responsible for the amazing growth in the power and the innovative potential of microchips."

Gordon Moore, a California native, had a degree in physics and chemistry from Caltech. Yet when he came to work for Shockley, he was barely able to define a semiconductor. He developed into a prophetic pioneer, perhaps best known for Moore's law. In 1965, for a special issue of *Electronics,* Moore reviewed the history of the increases in the number of transistors on each silicon chip. He then predicted that the number of chips would double each year. He later modified that to increase the time to eighteen months. But Moore's law has become the growth expectation in an industry whose growth seems boundless.

When they founded Intel, the plan was for the new company to make semiconductor chips to be used for memory storage. When a Japanese company, BUSICOM, asked them to make a chip that could also do calculations, Ted Hoff, at Intel, got the idea for a more complex and comprehensive product.

The early integrated circuits were all based upon the bipolar or junction transistor. In 1959, though, Bell Labs introduced a new type. This was the metal-oxide-semiconductor field-effect transistor (MOSFET), invented by Dawon Kahng and John Atalla. It was based on a design originally patented in 1925 and purchased by Bell Labs in order to protect the junction transistor. The first MOSFET models were not encouraging. They cost more and were less reliable than the junction transistors. They proved, however, to be simpler to manufacture and smaller than junction transistors.

In turn, Hoff based his design on MOSFET technology. Intel described its idea to BUSICOM: instead of the more basic design the Japanese engineers had conceptualized, this chip was to be an entire mini-computer. Hoff designed the chip, and engineer Frederico Faggin began building a workable model. BUSICOM, though, was experiencing some financial difficulties and became worried that Intel's idea was uncertain and was taking too long. Intel had confidence in its concept. It offered to return BUSICOM's $40,000 investment and take over ownership of the new chip. As Hoff described it:

> [BUSICOM wasn't] very interested in making any changes, the Japanese engineers were committed to the design they had and they essentially told me to stay

> out of their way. But Bob Noyce was encouraging, and so I continued to work on that, and over a relatively short period of time, felt that the way to do it would be to make a very simple, general purpose computer.

Noyce supervised nine months of development and design, and the company announced the creation of the world's first microprocessor—a single silicon chip that provided storage as well as the ability to program logic and arithmetic operations. That first chip, called the 4004, contained just over two thousand transistors etched into silicon—almost nothing by today's standards. Yet this tiny chip was as powerful as ENIAC, the early enormous computer built in 1946.

Improvements followed one after the other in quick succession. As Moore predicted, chips got smaller and faster and were able to do more and more. This invention has been described as "the most significant invention … since the discovery of the wheel and the development of written language." The story of semiconductors is still being written. And as Andrew Grove has said, "A fundamental rule in technology says that whatever can be done will be done."

Semiconductors are the reason that we live in a world of cell phones.

CHAPTER 5

The Influence of Semiconductors Today

If we want to find out about a new song, a famous person, what is happening in another country, or even what our friends have been doing, we have a lot of ways to get that information. We don't have to send smoke signals or learn how to read inscriptions carved on a stone or even read a book printed on paper in order to find out about what we want to know. We have an astonishing number of options for gathering and sharing data. And, in one word, the reason we have those options is semiconductors. In an era that is very brief when we think about the time periods in human history, semiconductors have gone from being a group of unusual and relatively unimportant substances to being at the very center of our modern world. The Information Age we live in is based on semiconductors and the technological advances that semiconductors have made possible. Yet the global changes that have come from semiconductors are still evolving, and the story of their influence is still being written.

COMPUTERS

The Apollo Guidance Computer

One of the first large-scale uses of the integrated circuit computer was in the Minuteman missile systems. The computer for missiles had to be fast, accurate, compact, and

The moon launch used semiconductor technology.

sturdy in order to withstand the thrust of a missile launch and still provide reliable flight calculations.

Technological development for military purposes, such as this one, improved quality control and supported development in circuit design. The Apollo Guidance Computer, created at MIT, reduced the size of the spacecraft's onboard computer from the equivalent of seven side-by-side refrigerators to one cubic foot. This computer guided the *Apollo 11* to the surface of the moon. Large-scale commercial computers were then designed using integrated circuits. But the greatest impact of the integrated circuit was yet to come with the introduction of personal computers.

Apple and the PC

In 1976, Steve Wozniak, a computer scientist from San Jose, California, founded Apple Computers, along with his friend Steve Jobs. Wozniak had an early fascination with electronics and was known for building devices from scratch. After quitting his job at Hewlett-Packard, Wozniak paired up with Jobs and, in Jobs's family's garage, worked to create a personal computer to compete with the computers that were being produced by IBM. While Wozniak was in charge of inventing the products, Jobs was in charge of marketing them.

Apple's debut invention was the Apple I. Their first order was for fifty systems from the Byte Shop in Mountain View, California. Total sales had amounted to about two hundred machines when, the following year, Apple announced its new Apple II. It was the first computer that included color graphics. This model sold in the millions and was a popular consumer computer for the next twenty years.

Wozniak was injured in a plane crash in 1981 and became less involved in the organization. Apple engineers worked on developing the Lisa computer, whose innovative graphical

Steve Jobs, along with this friend Steve Wozniak, founded Apple Computers.

user interface (GUI) later became the essential model for all computers. Meanwhile, IBM, the business giant, moved into the desktop computer field with its own personal computer, the Model 5150. This computer was the first to be designated as a "PC," and it gained wide commercial popularity. An entire business sector based on the manufacture of PC clones flourished, and it created an enormous industry for software, peripherals, and other PC accessories.

Apple continued working on its own models and, in 1984, introduced the Macintosh, the first computer that used a mouse for navigation. Many software packages were built in to the Macintosh, and its word-processing program was the first to feature WYSIWYG (what you see is what you get). In an ad that contained a not-too-subtle reference to IBM as the totalitarian Big Brother from the George Orwell book, the power of the Macintosh destroys the evil giant. The war between Macs and PCs was on and continues to this day.

Supercomputers

Personal computers continued to get faster, smaller, more powerful, and easier to use. But while the consumer computer was going in one direction, scientists began working with another idea—the supercomputer. In 1991, Caltech researchers began using the Intel Touchstone Delta, a machine with 512 independently operating processors arranged in a communication mesh. The supercomputer can perform intensive computations and facilitate activities such as real-time analysis of satellite pictures, molecular modeling for medical research, oil and gas exploration, climate research, and nuclear test simulation. The performance of supercomputers is being regularly increased and used for ever more varied applications.

NETWORKING

The invention and development of computers allowed us to store, process, and analyze information in ways that were never before possible. Semiconductors, and the microprocessors that were built using them, made it possible for an author to write a book without having to tear up rejected pages, for an accountant to easily keep the financial records for a business balanced and up to date, and for an engineer to calculate the stress factors that would influence the safety of a bridge under construction. Early computers, however, were still limited by the old ways of communicating all this new information. If the author or accountant or engineer wanted anyone else to see his or her information, it had to be delivered by mail or in person—often a slow and time-consuming process. Then came networking.

ARPANET

The idea of networking followed soon after the development of the computer. In 1962, J. C. R. Licklider of MIT wrote about a "Galactic Network" that would connect computers and allow someone to access data and programs from any one of the connected sites, much in the way that today's internet functions. Working at MIT on his doctoral thesis, Leonard Kleinrock developed the concept of "packet switching"—a theory of communication technology that became the basis for computer networking. Kleinrock wrote his first book on packet switching in 1964 and convinced his fellow researcher, Lawrence Roberts, that his theory could work. Roberts explored the idea of networking and connected a computer at MIT with a computer in California using a low-speed dial-up telephone. This was the first wide-area computer network. It proved that networking was possible, but it also proved that the telephone system was inadequate for the task.

In 1969, under the direction of the US Advanced Research Projects Agency (ARPA), four university computers were linked together using Kleinrock's packet-switching idea. This allowed data to be sent in small units, or "packets," that could be routed separately and reconstructed when they arrived at their destination. Packet-switching chips (integrated circuits that control the routing) make this system possible. The original goal was to connect scientific users so that they could share resources and information. Kleinrock's Network Measurement Center at UCLA (where he worked after graduating from MIT) was the first host computer connected in what was then known as ARPANET. The other hosts were Stanford Research Institute (SRI), UC Santa Barbara, and the University of Utah. When the first message was sent from UCLA to SRI, the internet was born. As Kleinrock described their attempts to transmit the word "login":

> At the UCLA end, they typed in the "l" and asked SRI if they received it; "got the l" came the voice reply. UCLA typed in the "o," asked if they got it, and received "got the o." UCLA then typed in the "g" and

> the darned system CRASHED! Quite a beginning. On the second attempt, it worked fine!

Additional computers were added to ARPANET, and the basic networking protocols were developed. By the mid-1970s, many small networks had been created. As they participated in ARPANET, it became a network of networks. The size of the ARPANET network continued to expand, and the Defense Communications Agency took charge of the program. In the 1980s, ARPANET was restructured and the military sites were separated from the other ARPANET facilities. By the mid-1980s there were ARPANET connections to networks in Europe, North America, and Australia. As ARPANET morphed into the internet, commercial internet service providers were increasing and internet users included commercial facilities as well as the original governmental, educational, and research organizations. ARPANET was retired in 1990. The "backbone" of the internet—the routes that local service providers use for long-distance connections—was turned over to a consortium of commercial providers.

World Wide Web

In 1994, Kleinrock chaired a National Research Council committee whose report, *Realizing the Information Future: The Internet and Beyond,* provided a vision for the internet as an open network serving all kinds of users, providing all kinds of information, from suppliers of all kinds. The report also noted the monumental impact that the internet exerted on a changing society:

> Networks are emerging as significant tools of social change, creating a new kind of free market for information services and connectivity, and extending the

Environmental Hazards

Semiconductor manufacturing is associated with a variety of environmental health and safety hazards, including exposure to highly corrosive chemicals such as hydrochloric acid; metals such as arsenic, cadmium, and lead; caustics such as ammonium hydroxide, hydrogen peroxide, and sodium hydroxide; volatile solvents such as methyl chloroform and acetone; and toxic gases such as arsine, phosphine, and silane.

It is difficult to obtain hard data about the results of that exposure. Semiconductor technology changes so quickly that old materials and processes are replaced before the health effects can be evaluated. Then, too, the industry can maintain confidentiality about its processes by claiming it is protecting industrial secrets.

Possible major health issues associated with working in a silicon manufacturing facility have included spontaneous abortions, an increase in urinary tract infections, lower white blood cell counts, and lung abnormalities. Cancer, though, is the most significant health risk, and cancer clusters have been identified in semiconductor industry workers and retired workers.

After a number of lawsuits alleging that environmental exposures had created illness in semiconductor workers, a series of studies were conducted, with contradictory and inconclusive results. There is clearly a need for more extensive research as well as ongoing promotion of preventative safety practices for what must be considered a high-risk employee group.

> boundaries for research and education in both time and space. The experiences of the research and education communities show that the Internet ... experiments are far from over—they have only just begun.

By this time, the internet had gone from the project of a small community of researchers to a broad cross section of commercial and personal users, and its revolutionary impact on society was becoming apparent. Coworkers in a single location could share data with one another or with colleagues across the country or across the world. The way people collaborated was changing. Email became the predominant form of business communication and a frequent choice for personal messages as well. The way people communicated was changing.

The World Wide Web continued to evolve, and users are now able to buy clothes, appliances, books, or even houses online. Social relationships are being reconstructed. People connect online and find partners, and online dating sites are consistently successful. Yet people can use the internet to avoid personal contact and remain isolated and disconnected from society.

Social networking has changed personal interactions; it has also changed political behavior and communication. Our ability to promote our personal opinions has been facilitated by the internet. The internet has also changed our relationship to information—we have access to more information but we can have difficulty evaluating it. E-medicine has made the internet a resource for medical information, diagnosis, and even treatment. Then, too, online education is making serious inroads into the on-campus enrollment at brick-and-mortar colleges and universities.

The internet is not only a way to communicate information, it is a forum for collaboration and interaction over great distances and boundaries. It is a stimulus for creating new ideas, an archive to honor old ones, and a mechanism for broadcasting those ideas to the world.

MOBILE SYSTEMS

Computers and the internet made it possible for us to be connected to everyone, but cell phones made it possible to be connected everywhere, all the time. Cellular technology, based on semiconductors, created the next great explosion in the Information Age. At the same time as cell phones were evolving, personal computers were getting smaller and smaller, making it possible to use them, too, as mobile devices.

When Alexander Graham Bell first created his photophone, the idea of a portable communication device seemed unlikely and unworkable. The first commercial mobile phones, based on analog technology, were large and unwieldy. But with the development and improvement of cellular technology, all that changed. Over just a few decades, improvements have been coming fast and furious, and the number of users—even in remote areas of the globe—has increased astronomically.

From Car Phones to Portability

Originally, mobile phones began as automobile systems. These systems consisted of a radio mounted in the car trunk with cables that connected it to a control unit mounted on the dashboard. Soon, satchel units were devised so that the phone was somewhat portable. Then Motorola created a handheld unit that weighed 2 pounds (0.9 kilogram) and could fit in a briefcase.

When cellular systems were first set up, they were partially digital and partially analog. Although much of the research developing these systems took place in the United States, Norway launched the first completely commercial system. As the systems became more popular, the technological limitations of the partially analog system became more obvious, particularly in the quality of service. By the late 1980s, however, voice-processing technology had made significant advances and second generation (2G) systems took advantage of these developments. Cell towers were able to handle many more transmissions, cutting down the cost to cellular carriers and making cell phones more accessible to consumers. With their increasing popularity, though, it soon became apparent that there was a need for greater data capability than the 2G system possessed.

The Push for Innovation

One of the developments providing the impetus for advances in the cell phone was the changes inaugurated in personal computers. In 1993, Apple joined the handheld computer field with its ill-fated personal data assistant (PDA) called Newton. Although the device contained a number of features, such as the handwriting recognition that would be incorporated in future handheld systems, it was not popular and was discontinued in 1998. Palm, Inc. introduced its own data organizer with the ability to connect to a PC or Mac computer and synchronize data on the two devices.

In 2002, the first camera phone was introduced by Japan's SoftBank. This JPhone integrated a camera with the digital phone and could share the photos wirelessly. New possibilities now emerged for the cell phone. It could be used for data transfer, downloading information from the internet, and sending video. This required a new capacity for data exchange with a higher speed capability. Although the 3G cell phones were in development, an interim solution was created with the 2.5 G devices—not all the options expected with 3G, but a medium speed data transfer that could provide some desired innovations.

While there were beginning to be some overlapping functions between PDAs and cell phones, cell phones were still basically used for making calls, and PDAs could store a contact list, organize your activities, and synchronize with your computer. But the overlapping became more widespread. The union of cell phones and PDAs eventually became a smartphone.

Smartphones, Tablets, and More

The first commercially released device that could be thought of as a smartphone was the Simon Personal Communicator, released by Bell South in 1994. The Simon, with a touch-screen display, could make and receive calls. It could also handle faxes and emails and provide organization through an address book, calendar, scheduler, notepad, and calculator. The Ericson R380 was the first device sold as a "smartphone."

In 2007, Apple introduced the iPhone. This mobile device could download information from the internet, take high-definition pictures, send text messages, organize your calendar, and play music. Almost incidentally, it could also make and receive phone calls. Other iPhone features included GPS navigation and voice dictation. A variety of apps, available from the App Store, could provide additional functionality. Originally, the iPhone was exclusively available for the AT&T cellular network.

In 2008, the Android, developed by Google and included on devices manufactured by Samsung, was released. The Android operating system was intended for cell phones and tablet computers. The Android is an open source system with a community of programmers contributing to its development and modification rather than being a product of a single organization. Google promised to support the software through providing a flexible system with regular upgrades. Like the iPhone, Android cell phones have numerous apps available to enhance their capabilities.

In 2012, Apple released its iPad, a device with many iPhone features but with a 9-inch (22.86-cm) screen and no phone. The iPad, with its small size but convenient screen has become popular, in part because of apps that support many creative activities: making movies, creating art, and music production. Android-based tablets have also provided a significant portion of the tablet market sales, and Windows-based systems have made recent inroads.

Apple introduced the Apple Watch in 2015 and, in a market full of innovation, achieved a new buzz level by creating an often attempted but never successful dream: a computer in the form of a watch. This product, compatible with iPhones and MacBooks, includes sensors for environmental and health monitoring. On the day of its release, over one million watches were sold.

The Range of Mobile Devices

It's obvious when you walk down the street that cell phones are everywhere. Cellular technology has allowed people to have phones where it had never before been possible. In some places, there are more cell phones than people. Landlines are being steadily replaced by cell phones. And we can communicate instantly with people in places inaccessible in other ways. Seniors are learning how to text in order to connect to their grandchildren, and young children carry cell phones so that their parents can assure themselves of their child's safety.

Business is conducted at all hours and in all places because the cell phone makes employees always accessible. The cell phone has contributed to events that may be frivolous or consequential, from flash mobs to political revolutions. Yet the sale of cell phones also has a significant economic influence.

The cell phone is certainly one of the most powerful components of the Information Age because it makes advanced technology readily available and it provides global information, global products, and global perspectives to all who get connected.

The INTERNET of THINGS

The "Internet of Things" (IoT) is a phrase that refers to a new direction for communications. People have been connected to the internet—now things are getting connected, too. The

The Internet of Things links electronic devices and allows them to communicate.

connection is providing a wide range of new possibilities and will provide even more in the future. Smartwatches are a simple example. Connecting information gathered by sensors on the watch with data on the internet can provide, for example, advice on how the watch wearer might modify sleep patterns to avoid insomnia. Another possibility could be if sensors on a milk carton in a smart refrigerator were able to notify a homeowner when the milk was getting low and it was time to buy more.

The basic concept uses sensors that measure data and cloud-based applications that interpret and synthesize the information that is being transmitted. The impact that the IoT can have on our lives can be minimal or monumental. It could be paper towel dispensers signaling the need for a refill or sensors that send a patient's vital signs in real time to a health care provider so that any necessary medical interventions can be undertaken quickly. It might be an oven that could be turned on, or a bathtub filled, or a coffee maker started remotely, from the user's phone. It could also be sensors that alert users to danger like fire, or moisture, or a break-in when they are away from home.

The IoT involves small systems like a household, larger systems like a business, and much larger systems, such as those involved in agricultural production. Smart appliances in the home could schedule their activity, like dishwashing, for off-peak times. General Electric (GE) has developed a range of sensors and software to help companies function more efficiently. For example, GE has helped Air Asia manage its fuel usage, and the company expects to save between $30 million to $50 million in costs over a five-year period. Monsanto and other agribusiness giants are using IoT technology to improve planting and harvesting schedules and procedures. The *New York Times* reported that IoT agriculture involves:

> Exploiting data from many sources—sensors on farm equipment and plants, satellite images and weather tracking. In the near future, the use of water and fertilizer will be measured and monitored in detail, sometimes on a plant-by-plant basis.

Monsanto has been using data collected from combines to measure crop yields, moisture, and soil quality in order to get higher production. It also tracks the location and temperature of seed shipments to modify its route schedule and minimize seed loss.

Self-Driving Cars

Some of the IoT applications seem like science fiction but are available now. Although the driverless car is a vision for the future, semiautonomous automobiles already perform some tasks that will be incorporated in the vehicles currently being designed. Some cars can park themselves, apply the brakes to avoid crashes, and even stay in the correct lane on the highway. But the completely autonomous car is still in the works.

Google, a major player in the development of the self-driving car, is already in the third generation of design, and its vehicles have logged more than one million driverless miles. It announced a plan for a car with no pedals and no steering wheels in May 2014. It unveiled its first functioning prototype seven months later. Other companies have also ventured into the driverless car market. An Audi equipped with IoT software traveled almost 3,500 miles (5,632 km) without a human driver. Volvo, too, plans, to have driverless vehicles used on a short stretch of roadway in Sweden. Microprocessing chip pioneer Intel has paired with BMW, the German car manufacturer, and Mobileye, an Israeli technology company, to produce driverless cars. Mobileye's EyeQ chips are already used to process images providing cars with locational awareness, and the fifth generation of these devices will be used in the new cars. Although we are not quite at the era of the driverless car, the National Highway Traffic Safety Administration (NHTSA), the US governmental agency in charge of safety standards, has declared that there is a driver of an autonomous car: the software that controls it.

The autonomous car may not be here yet, but there is already speculation about its impact, particularly on the lives of the elderly, who may be facing a loss of independence with their decreasing driving skills. The driverless car is being promoted as a way to give some seniors a new lease on life. If the technology is made truly accessible, older citizens will have a way to physically stay involved in community activities, provide volunteer services, and perhaps even continue working.

More IoT Possibilities

The IoT is currently serving a variety of functions and is expected to serve a huge role in the future. Roadways, bridges, dams, and other essential infrastructure elements can be equipped with sensors detecting damage and deterioration and provide alerts for any necessary maintenance or upgrades. Manufacturers can use IoT programs to improve efficiency, coordinate activities, and find cost-saving and quality-control methods to enhance the business. In the smart cities of the future, traffic sensors and GPS information will make traffic flow more smoothly and provide alternate routes for harried commuters.

Health care has already been transformed by the IoT. Smartwatches and fitness trackers are well known, but medical-grade sensors are moving onto the market. Software from Microsoft analyzes the data from a wearable device that provides real-time patient monitoring. This can save lives and time, and improve diagnoses. A Google subsidiary is working with a pharmaceutical firm to develop a bioelectronics company aiming "to tackle a wide range of chronic diseases using miniaturized, implantable devices that can modify electrical signals that pass along nerves in the body, including irregular or altered impulses that occur in many illnesses." Researchers at the University of Michigan are developing a sensor that can detect a very low level of cancer cells, as low as three to five cancer cells in one milliliter in a blood sample, and provide extraordinarily early detection capabilities.

Cisco, a technology conglomerate, forecasts that by 2020 there will be over fifty million IoT devices and that, by 2025, the IoT market will be worth $19 trillion. What may sound like an inevitable positive trajectory for the IoT does, however, have critics who present a cautionary note. Justin Reich, a fellow at Harvard University's Berkman Center for Internet and Society said:

> It will have widespread beneficial effects, along with widespread negative effects. There will be conveniences and privacy violations. There will be new ways for people to connect, as well as new pathways towards isolation, misanthropy, and depression. I'm not sure that moving computers from people's pockets (smartphones) to people's hands or face will have the same level of impact that the smartphone has had, but things will trend in the similar direction. Everything that you love and hate about smartphones will be more so.

SEMICONDUCTORS, A CRITICAL COMPONENT

Reviewing the impact that semiconductors have had, through their uses in computers, networking, mobile devices, and the Internet of Things, provides a broad but incomplete picture of the ways in which semiconductors have influenced how we live today. Less than one hundred years ago, we hardly had any awareness of semiconductors. Now they are part of virtually everything we say and do. Although we have looked at some of the major applications of the semiconductor, the list grows larger each minute: solar panels, fiber optics, and nanomedicine are just a few significant examples. It would be difficult to find much in our human history that so clearly has had the tremendous impact that these peculiar substances have exerted. We talk differently, we socialize differently, we do business differently, and we connect differently than we did before the era of semiconductors. And our world, under the influence of the semiconductor technology, continues to change at a rapid rate.

Chronology

1791 Michael Faraday is born

1831 James Clerk Maxwell is born

1833 Michael Faraday records observation of first semiconductor effect

1856 J. J. Thomson is born

1862 James Clerk Maxwell publishes Maxwell's equations

1871 Frederic Braun discovers semiconductor point-contact rectifier effect

1880 Alexander Graham Bell invents the photophone

1897 J. J. Thomson discovers the electron

1901 Jagadish Bose patents the cat's whisker crystal rectifier device

1924 Jean Hoerni is born

1931 Alan Wilson publishes *The Theory of Electronic Semiconductors*

1940 Russell Ohl discovers the p-n junction and photovoltaic effects in silicon

1947 John Bardeen and Walter Brattain invent the point-contact transistor

1948 William Shockley invents the junction transistor

1955 Steve Jobs is born

1959 Jean Hoerni develops the planar process for manufacturing semiconductor chips

1960 Robert Noyce and Jack Kilby independently develop integrated circuits

1971 Intel's microprocessor goes public

1999 The term "Internet of Things" is first used

2007 Apple debuts iPhone

2016 Google creates an independent company, Waymo, to continue work on driverless cars

Glossary

alternating current An electric current that begins flowing in one direction and then regularly reverses course.

Ampere's law One of Maxwell's equations (not to be confused with Ampere's law on force) that determines the magnetic field associated with a given current, or the current associated with a specific magnetic field.

amplifier A device that increases a signal.

capacitor Also known as a condenser, a device for accumulating and storing an electric charge.

cathode An electrode with a negative charge.

cathode ray A beam of electrons emitted from the cathode of a high-vacuum tube.

diode An electronic component with two electrodes that allows current to move in one direction.

direct current An electric current that flows one way.

doping Adding a small number of atoms from another material to a pure semiconductor in order to control its conductivity.

electrical current Measured in amperes, the rate an electric charge moves.

electrode An electrical conductor used to make contact with a nonmetallic part of a circuit (e.g., a semiconductor, an electrolyte, a vacuum, or air).

electrolysis Process by which electric current is passed through a substance to effect a chemical change.

electromagnetism The science of charge and of the forces and fields associated with charge.

electron A negatively charged subatomic particle. It can be either free (not attached to any atom) or bound to the nucleus of an atom. In a semiconductor, it is, along with a hole, one of two possible charge carriers.

electroscope An early scientific device used to test whether or not an object has an electric charge.

fiber optics Technology that uses glass threads (fibers) to transmit information. A fiber-optic cable consists of a bundle of glass threads, each of which is capable of transmitting messages that are communicated using light waves.

germanium A chemical element that is a semiconductor.

hole An electric charge carrier with a positive charge.

integrated circuit A small electronic device made of semiconductor materials that can hold anywhere from hundreds to millions of electronic components, including transistors and capacitors.

ion An electrically charged atom or group of atoms formed by the loss or gain of one or more electrons.

junction transistor A transistor that relies on contact with three parts: the emitter, the base, and a collector.

microprocessor An integrated circuit that contains the central processing unit for a computer.

nanoparticle Any particle less than 100 nanometers; a nanometer is one-billionth of a meter.

n-type semiconductor A semiconductor in which conductivity is due primarily to the movement of electrons.

photoelectron An electron emitted when light comes into contact with a material that is sensitive to light.

photovoltage The electric force developed by a device that is sensitive to light.

planar process A silicon transistor manufacturing process.

point-contact rectifier A device that converts alternating current to direct current by placing a thin metal wire in contact with a semiconductor point-contact transistor (a transistor that has a base electrode and two or more point contacts).

p-type semiconductor A semiconductor in which the conductivity is due primarily to the movement of positive holes.

quantum theory The theoretical basis of modern physics that addresses the nature and behavior of matter and energy on the atomic and subatomic level.

rectification The process of converting an alternating current to a direct current.

resistor A component that puts electrical resistance in a circuit.

semiconductor An important part of most modern circuits; a solid material (like germanium or silicon) that is neither a conductor nor an insulator; a material that is more conductive as it is exposed to light or an electric field or as its temperature increases.

silicon The most abundant electropositive element in Earth's crust and a critical semiconductor used in electronic components.

solid-state electronics Components that control current without the use of a moving parts transistor (a device that regulates voltage flow and acts as a switch or gate for electric current).

transistor A device that modifies or directs the flow of energy using a semiconductor.

vacuum tube A tube from which most or all of the air has been pumped that may contain a number of electrodes.

Further Information

BOOKS

Berlin, Leslie. *The Man Behind the Microchip: Robert Noyce and the Invention of Silicon Valley*. New York, NY: Oxford University Press, 2005.

Levinshtein, M. E., and G. S. Simin. *Getting to Know Semiconductors*. Singapore: World Scientific, 1992.

WEBSITES

The Discovery of the Electron
http://history.aip.org/history/exhibits/electron/jjhome.htm

Learn more about this important scientific advance from the Center for the History of Physics, a part of the American Institute of Physics.

Semiconductors
http://www.nobelprize.org/educational/physics/semiconductors/1.html

This site is a great explanation from Nobelprize.org of the science of semiconductors.

VIDEOS

"Absolute Genius | Michael Faraday | S1E4"
https://www.youtube.com/watch?v=QyrNh4UXrWw

Absolute Genius hosts Dick and Dom provide a lighthearted look at Michael Faraday's groundbreaking work.

"Build a Photophone"
https://www.youtube.com/watch?v=fslMMQX7tfQ

Learn more about Alexander Graham Bell's great unknown invention: a mobile phone.

"How Michael Faraday (1791–1867) Shed New Light on Electrochemistry"
https://www.youtube.com/watch?v=qVhiwi6AvQM

Professor Gary Patterson explains Faraday's contributions.

"What Is a Semiconductor?"
https://www.youtube.com/watch?v=gUmDVe6C-BU

An MIT student describes semiconductors and demonstrates their properties.

Bibliography

Anderson, Janna, and Lee Rainie. "The Internet of Things Will Thrive by 2025." Pew Research Center. Retrieved November 22, 2016. http://www.pewinternet.org/2014/05/14/internet-of-things.

Baggot, Jim. "The Myth of Michael Faraday." *New Scientist*, September 21, 1991. https://www.newscientist.com/article/mg13117874-600-the-myth-of-michael-faraday-michael-faraday-was-not-just-one-of-britains-greatest-experimenters-a-closer-look-at-the-man-and-his-work-reveals-that-he-was-also-a-clever-theoretician/.

Barbour, Eric. "How Vacuum Tubes Work." Vacuumtubes.net. Retrieved November 5, 2016. http://www.vacuumtubes.net/How_Vacuum_Tubes_Work.htm.

Berlin, Leslie, and H. Casey. "Robert Noyce and the Tunnel Diode." *IEEE Spectrum*, May 2, 2005. http://spectrum.ieee.org/biomedical/devices/robert-noyce-and-the-tunnel-diode.

Biography.com. "Johannes Gutenberg: Biography." Retrieved November 1, 2016. http://www.biography.com/people/johannes-gutenberg-9323828.

———. "Steve Wozniak Biography." Retrieved November 15, 2016. http://www.biography.com/people/steve-wozniak-9537334#synopsis.

———. "William B. Shockley Biography." Retrieved October 29, 2016. http://www.biography.com/people/william-b-shockley-9482432.

Bragg, Melvyn. "Conductors and Semiconductors." *BBC4: In Our Time,* February 23, 2012. http://www.bbc.co.uk/programmes/b01c7sml.

Braun, Ernest, and Stuart MacDonald. *Revolution in Miniature.* London, UK: Cambridge University Press, 1982.

Briggs, Josh. "Who Invented the Radio?" How Stuff Works: Science. Retrieved November 3, 2016. http://science.howstuffworks.com/innovation/inventions/who-invented-the-radio.htm.

Chemical Heritage Foundation. "Joseph John Thomson." Retrieved November 2, 2016. https://www.chemheritage.org/historical-profile/joseph-john-%E2%80%9Cj-j%E2%80%9D-thomson.

Complete Dictionary of Scientific Biography. "Alan Herries Wilson." Encyclopedia.com. Retrieved November 17, 2016. http://www.encyclopedia.com/science/dictionaries-thesauruses-pictures-and-press-releases/wilson-alan-herries.

Edison Tech Center. “Semiconductors.” Retrieved November 5, 2016. http://www.edisontechcenter.org/semiconductors.html.

Encyclopedia Britannica. “Erwin Schroeder.” Retrieved November 3, 2016. https://www.britannica.com/biography/Erwin-Schrodinger.

———. “Louis de Broglie.” Retrieved November 4, 2016. https://www.britannica.com/biography/Louis-de-Broglie.

———. “Robert Noyce, American Engineer.” Retrieved November 1, 2016. https://www.britannica.com/biography/Robert-Noyce.

———. “Sir J. J. Thomson, British Physicist.” Retrieved November 2, 2016. https://www.britannica.com/biography/J-J-Thomson.

Famous Scientists. “Ernest Rutherford.” Famousscientists.org, July 30, 2015. http://www.famousscientists.org /ernest-rutherford.

———. “James Clerk Maxwell.” Famousscientists.org, July 1, 2014. http://www.famousscientists.org/james-clerk-maxwell.

———. “J. J. Thomson.” Famousscientists.org, July 27, 2015. http://www.famousscientists.org/j-j-thomson.

Gladstone, J. H. *Michael Faraday*. London, UK: Macmillan, 1872.

History.com. "Alexander Graham Bell." Retrieved November 3, 2016. http://www.history.com/topics/inventions/alexander-graham-bell.

———. "Guglielmo Marconi." Retrieved November 3, 2016. http://www.history.com/topics/inventions/guglielmo-marconi.

History of Computing Project. "Walter Brattain." Retrieved November 15, 2016. http://www.thocp.net/biographies/brattain_walter.htm.

IEEE Computer Society. "Jean Hoerni." Retrieved November 4, 2016. https://www.computer.org/web/awards/mcdowell-jean-hoerni.

Institution of Engineering and Technology. "Archives Biographies: Michael Faraday." Retrieved November 1, 2016. http://www.theiet.org/resources/library/archives/biographies/faraday.cfm.

Intel Newsroom. "Robert Noyce." Retrieved November 1, 2016. https://newsroom.intel.com/biography/robert-noyce/.

Langlois, Richard N. "Computers and Semiconductors." Retrieved November 5, 2016. http://web.uconn.edu/ciom/DTEG.pdf.

Lecuyer, Christophe. *Making Silicon Valley*. Cambridge, MA: MIT Press, 2006.

Neiger, Chris. "Virtual Reality vs. the Internet of Things: Which Tech Trend Has More Opportunity?" The Motley Fool, October 25, 2016. http://www.fool.com/investing/2016/10/25/virtual-reality-vs-the-internet-of-things-which-te.aspx?source=djc.

Nobelprize.org. "John Bardeen – Biographical." Retrieved November 7, 2016. http://www.nobelprize.org/nobel_prizes/physics/laureates/1956/bardeen-bio.html.

———. "Niels Bohr – Biographical." Retrieved November 7, 2016. http://www.nobelprize.org/nobel_prizes/physics/laureates/1922/bohr-bio.html.

———. "William B. Shockley – Biographical." Retrieved November 6, 2016. http://www.nobelprize.org/nobel_prizes/physics/laureates/1956/shockley-bio.html.

Orton, John. *The Story of Semiconductors*. Oxford, UK: Oxford University Press, 2004.

PBS. "The Invention of the Microprocessor." Retrieved November 1, 2016. http://www.pbs.org/transistor/background1/events/micropinv.html.

———. "Robert Noyce." Retrieved November 1, 2016. http://www.pbs.org/transistor/album1/addlbios/noyce.html.

———. "Russell Ohl." Retrieved November 14, 2016. http://www.pbs.org/transistor/album1/ohl/index.html.

———. "William Shockley." Retrieved November 2, 2016. http://www.pbs.org/transistor/album1/shockley/.

Riordan, Michael. "The Silicon Dioxide Solution." IEEE Spectrum, December 1, 2007. http://spectrum.ieee.org/semiconductors/design/the-silicon-dioxide-solution.

Royal Institution. "Michael Faraday in London." Retrieved November 2016. http://www.rigb.org/our-history/michael-faraday/faraday-walk.

Saxon, Wolfgang. "William B. Shockley, 79, Creator of Transistor and Theory on Race." *New York Times,* August 14, 1989. http://www.nytimes.com/learning/general/onthisday/bday/0213.html.

Scott, David Clark, "Robert Noyce: Why Steve Jobs Idolized Noyce." *Christian Science Monitor,* December 12, 2011. http://www.csmonitor.com/Technology/2011/1212/Robert-Noyce-Why-Steve-Jobs-idolized-Noyce.

Silicon Engine. "The First Semiconductor Effect Is Recorded." Retrieved November 2, 2016. http://www.computerhistory.org/siliconengine/first-semiconductor-effect-is-recorded.

Smith, Willoughby. "Letter to Latimer Clark, February 4, 1873." Retrieved November 6, 2016. http://histv2.free.fr/selenium/smith.htm.

Stewart, William. "ARPANET – The First Internet." Living Internet. Retrieved November 15, 2016. http://www.livinginternet.com/i/ii_arpanet.htm.

Yu, Albert. *Creating the Digital Future.* New York, NY: The Free Press, 1998.

Index

Page numbers in **boldface** are illustrations. Entries in **boldface** are glossary terms.

About the Author

Grace Murphy has written books, white papers, speeches, encyclopedia articles, and online courses in history, economics, and science. Murphy has also been employed as a cook, educational consultant, health care administrator, computer programmer, college teacher, and clinical social worker/ psychotherapist. She currently lives in the peaceful hamlet of Shady, New York, where she enjoys regular hikes on the local mountain trails.